Nanomaterials
for Photocatalytic Chemistry

World Scientific Series in Nanoscience and Nanotechnology*

ISSN: 2301-301X

Series Editor-in-Chief: Frans Spaepen (*Harvard University, USA*)

Published

Vol. 7   Scanning Probe Microscopy for Energy Research:
Materials, Devices, and Applications
*edited by Dawn A. Bonnell (The University of Pennsylvania, USA)
and Sergei V. Kalinin (Oak Ridge National Laboratory, USA)*

Vol. 8   Polyoxometalate Chemistry: Some Recent Trends
edited by Francis Secheresse (Université de Versailles-St Quentin, France)

Vol. 9   Handbook of Biomimetics and Bioinspiration: Biologically-Driven Engineering of Materials, Processes, Devices, and Systems
*edited by Esmaiel Jabbari (University of South carolina, USA),
Deok-Ho Kim (University of Washington, USA),
Luke P. Lee (University of California, Berkeley, USA),
Amir Ghaemmaghami (University of Nottingham, UK) and
Ali Khademhosseini (Harvard University, USA & Massachusetts Institute of Technology, USA)*

Vol. 10 Pore Scale Phenomena: Frontiers in Energy and Environment
*edited by John Poate (Colorado School of Mines, USA),
Tissa Illangasekare (Colorado School of Mines, USA),
Hossein Kazemi (Colorado School of Mines, USA) and
Robert Kee (Colorado School of Mines, USA)*

Vol. 11 Molecular Bioelectronics: The 19 Years of Progress (2nd Edition)
by Nicolini Claudio (University of Genoa, Italy)

Vol. 12 Nanomaterials for Photocatalytic Chemistry
edited by Yugang Sun (Temple University, USA)

*For the complete list of volumes in this series, please visit
www.worldscientific.com/series/wssnn

Volume **12**

World Scientific Series in
Nanoscience and Nanotechnology

Nanomaterials
for Photocatalytic Chemistry

Editor

Yugang Sun

Temple University, USA

NEW JERSEY · LONDON · SINGAPORE · BEIJING · SHANGHAI · HONG KONG · TAIPEI · CHENNAI · TOKYO

Published by

World Scientific Publishing Co. Pte. Ltd.
5 Toh Tuck Link, Singapore 596224
USA office: 27 Warren Street, Suite 401-402, Hackensack, NJ 07601
UK office: 57 Shelton Street, Covent Garden, London WC2H 9HE

Library of Congress Cataloging-in-Publication Data
Names: Sun, Yugang, editor.
Title: Nanomaterials for photocatalytic chemistry / editor, Yugang Sun (Temple University, USA).
Other titles: World Scientific series in nanoscience and nanotechnology ; v. 12.
Description: Hackensack, NJ : World Scientific, [2016] | Series: World Scientific series in
  nanoscience and nanotechnology ; volume 12 | Includes bibliographical references and index.
Identifiers: LCCN 2016029078 (print) | ISBN 9789813141995 (hardcover ; alk. paper) |
  ISBN 9813141999 (hardcover ; alk. paper)
Subjects: LCSH: Nanostructured materials--Industrial applications. | Photocatalysis. | Solar energy.
Classification: LCC TA418.9.N35 N25925 2016 (print) | DDC 541/.395--dc23
LC record available at https://lccn.loc.gov/2016029078

British Library Cataloguing-in-Publication Data
A catalogue record for this book is available from the British Library.

Copyright © 2017 by World Scientific Publishing Co. Pte. Ltd.

All rights reserved. This book, or parts thereof, may not be reproduced in any form or by any means, electronic or mechanical, including photocopying, recording or any information storage and retrieval system now known or to be invented, without written permission from the publisher.

For photocopying of material in this volume, please pay a copying fee through the Copyright Clearance Center, Inc., 222 Rosewood Drive, Danvers, MA 01923, USA. In this case permission to photocopy is not required from the publisher.

Desk Editor: Rhaimie Wahap

Typeset by Stallion Press
Email: enquiries@stallionpress.com

Printed in Singapore

Preface

Solar energy represents the most important source of renewable energy, which can be harnessed using a range of ever-changing technologies including solar heating, photovoltaics, solar thermal energy, solar architecture and artificial photosynthesis. Many of these technologies have been well developed and broadly delivered to serve our society. This book focuses on the emerging area in utilization of photon energy for catalyzing useful chemical reactions (also called artificial photosynthesis), ranging from materials design at nanometer scale to nanomaterials synthesis to photocatalytically chemical conversion. Chapter 1 presents a special design for the conversion of earth-abundant white TiO_2, which can only absorb ultraviolet (UV) light, to black TiO_2 nanomaterials, enabling effective absorption of visible light and thus, enhanced photocatalytic activities. Chapter 2 summarizes the use of nanomaterials of rust (e.g. hematite α-Fe_2O_3) and derivatives as photoanodes in photoelectrochemical (PEC) cells to drive water splitting reaction, which can produce hydrogen fuel from water, by harnessing solar energy. Advances in design of both the nanostructures and the system level of PEC cells are comprehensively overviewed. Nanowires of silicon (the second most earth-abundant element) are discussed in Chapter 3, shedding light on the importance of heterogeneous junctions in determining performance of photocatalysis. In Chapter 4, composite nanomaterials made of optically active components and graphene sheets are discussed to highlight their improved photocatalytic activities towards selective organic transformations in comparison with their single component counterparts. In addition to semiconductor nanomaterials, noble metal nanoparticles that exhibit strong surface plasmon resonances, show the promise in photocatalysis because excitation of plasmons can generate highly energetic hot electrons to drive reduction reactions directly on metal nanoparticle surface, or to migrate to the conduction band of adjacent semiconducting domains to promote reduction reactions. The fundamental principles and examples in various photocatalytic chemical reactions such as selective epoxidation, selective alcohol oxidation, coupling reactions, etc. are reviewed in Chapters 5 and 6. Relaxation of the excited plasmons eventually results in localized heating of the metallic nanostructures, and the corresponding thermal energy is also able to drive certain chemical reactions with improved selectivity. In Chapter 7, using natural biological chromophores for harvesting solar energy is discussed to highlight their capability and efficiency in photocatalytically driving chemical reactions when they form hybrid structures with inorganic nanostructures. The last chapter [Chapter 8] comprehensively reviews the use of photocatalytically active nanomaterials to facilitate reduction of CO_2 into useful chemicals.

Finally, I would like to thank all of the contributing authors, who are from both academia and national laboratory, for their time and effort in preparing the manuscripts presented in this book. Their work, and that of others in this emerging area are essential to the vibrancy and fast pace of progress in this field. I hope that this book can serve a useful reference for those new to this field of research or already engaged in it, from graduate students to postdoctoral fellows and practicing researchers.

Yugang Sun
Temple University
June 2016

Contents

Preface   v

Chapter 1   Black Titanium Dioxide (TiO_2) Nanomaterials   1
Xiaodong Yan, Lihong Tian and Xiaobo Chen

Chapter 2   Efficient Photocatalysis using Hematite Nanostructures and their Derivatives   27
Chun Du, James E. Thorne and Dunwei Wang

Chapter 3   One-Dimensional Silicon Nanowire Composites for Photocatalysis   57
Yuanyuan Ma, Jiayuan Li and Yongquan Qu

Chapter 4   The Applications of Graphene-based Nanocomposites in the Field of Photocatalytic Selective Organic Transformations   81
Min-Quan Yang, Siqi Liu, Bo Weng and Yi-Jun Xu

Chapter 5   Plasmonic Photocatalysts   117
Congjun Wang and Christopher Matranga

Chapter 6	Plasmon-assisted Chemical Reactions *Ruibin Jiang and Jianfang Wang*	155
Chapter 7	Harnessing Nature's Purple Solar Panels for Photoenergy Conversion *Elena A. Rozhkova and Peng Wang*	195
Chapter 8	Status and Perspectives on the Photocatalytic Reduction of CO_2 *Jun Zhang and Zhaojie Wang*	229
Index		289

Chapter 1

Black Titanium Dioxide (TiO$_2$) Nanomaterials

Xiaodong Yan[*], Lihong Tian[*,†] and Xiaobo Chen[*,‡]

[*]Department of Chemistry, University of Missouri–Kansas City
Kansas City, MO 64110, USA
[†]Hubei Collaborative Innovation Center for Advanced
Organic Chemical Materials, Hubei University Wuhan,
Hubei 430062, PRC
[‡]chenxiaobo@umkc.edu

Over the past decades, titanium dioxide (TiO$_2$) nanomaterials have attracted tremendous attention in a large variety of applications, e.g. photocatalytic water splitting and environmental pollution removal. However, the large bandgap of TiO$_2$ nanomaterials restricts their optical absorption to ultraviolet light. The discovery of black TiO$_2$ nanomaterials by hydrogenation has extended their optical absorption beyond the entire visible spectrum and greatly enhanced their photocatalytic activities. Herein, the progress in black TiO$_2$ nanomaterials has been reviewed with special emphasis on their synthesis methods along with their various chemical/physical properties and applications.

1. Introduction

Titanium dioxide (TiO_2) is well known for its photocatalytic activity towards water splitting and environmental pollution removal.[1-5] Its photocatalytic efficiency is highly dependent on the ability to absorb light in generating effective charge carriers (excited electrons in the conduction band and holes in the valence band).[6,7] The amount of charge carriers depends on the amount of light absorbed. Generally, the more light TiO_2 absorbs, the more charge carriers can be generated. Regardless of various crystal phases (e.g. anatase, rutile and brookite),[2] TiO_2 has a large electronic bandgap of 3.0–3.2 eV.[1,7] This limits its optical absorption in the ultraviolet (UV) spectrum. Unfortunately, UV light only accounts for ~5% of the solar light.[8] This limits the overall efficiency of TiO_2 in utilizing solar energy for photocatalytic water splitting and environmental pollution removal. Therefore, extending the optical absorption of TiO_2 is highly desired. To date, visible-light absorption of TiO_2 has been achieved by doping various metal ions and/or non-metal elements,[9-11] where the metal ions replace the Ti^{4+} ions and non-metal elements replace the O^{2-} ions in the TiO_2 lattices. Meanwhile, visible-light absorption has also been obtained by introducing intrinsic defects, such as oxygen vacancy and Ti^{3+} ions.[12] In 2011, extended absorption of TiO_2 beyond the entire visible-light region was made with the breakthrough of black TiO_2 nanoparticles through hydrogenation treatment.[13] The black TiO_2 nanoparticles possess a largely narrowed bandgap of ~1.5 eV with an optical onset of ~1.0 eV.

Herein, we would like to review the recent progress on black TiO_2 nanomaterials. We mainly focus on the synthesis of black TiO_2 nanomaterials, along with a brief introduction of their chemical and physical properties and their various applications. It should be noted that black TiO_2 nanomaterials cannot guarantee excellent performance in all applications, though they have witnessed great success in photocatalytic area. However, black TiO_2 nanomaterials achieved by introducing heteroatoms lies beyond the scope of this chapter.

2. Synthesis of Black TiO_2 Nanomaterials

2.1 Hydrogenation treatment

Since black TiO_2 nanoparticles were created by hydrogenation in 2011,[13] hydrogenation treatment has been the most widely used method to achieve black TiO_2 nanomaterials. Hydrogenation here refers to a chemical reaction with molecular hydrogen (H_2). It is very natural to make the assumption that hydrogen atoms will be attached to the TiO_2 through the formation of Ti–H bond, O–H bond or both during the hydrogenation process. This can greatly change the structure of TiO_2 nanomaterials, especially the outer layer of the TiO_2 crystals.[13–16] Generally, black TiO_2 nanomaterials from different research groups presented different properties and performance, owing to the variations in the preparation and hydrogenation processes.[17] However, these variations and complexities offer versatility for tailoring the surface structure of TiO_2 nanomaterials by controlling the preparation and hydrogenation processes. In this section, we thus critically examine the recent synthetic approaches of black TiO_2 nanomaterials and highlight the effect of hydrogenation condition on the diversity of hydrogenated TiO_2.

2.1.1 Hydrogenation in pure hydrogen atmosphere

High-pressure hydrogenation: Chen *et al.* reported the preparation of black anatase TiO_2 nanoparticles by treating pristine white TiO_2 nanoparticles in a vacuum for 1 h and then in 20.0 bar H_2 atmosphere at about 200°C for five days.[13] The pristine TiO_2 nanoparticles were obtained by heating the reaction solution containing titanium tetraisopropoxide, Pluronic F127, hydrochloric acid, deionized water and ethanol with the molar ratio of 1:0.005:0.5:15:40 at 40°C for 24 h, followed by evaporating and drying at 110°C for 24 h and calcinating at 500°C for 6 h.[13] Figure 1 shows the high resolution transmission electron microscopy (HRTEM) images of (a) pristine white TiO_2 nanoparticles and (b) black hydrogenated TiO_2 nanoparticles.[13] The black TiO_2 nanoparticle featured a core/shell structure, where the core presented ordered crystal lattice and

Fig. 1. HRTEM images of TiO$_2$ nanocrystals (a) before and (b) after hydrogenation (the insets in (a) and (b) are the digital images of the corresponding samples).[13] Reproduced from Ref. 13. Copyright 2011, The American Association for the Advancement of Science.

the shell was lattice-disordered.[13–16] The disordered shell was believed to host the possible hydrogen dopant in the form of Ti–H and O–H bonds, bringing about the midgap states and the black color of the hydrogenated TiO$_2$ nanoparticles.[13–16]

The color change of TiO$_2$ nanomaterials during the hydrogenation process was related to the exposed facets of TiO$_2$ nanocrystals and the amount of hydrogen incorporated onto the surface. The hydrogen incorporation and storage were different in well-defined nanocrystals of anatase TiO$_2$ with predominant (001) or (101) surface termination.[18] The (001)- and (101)-faceted TiO$_2$ nanocrystals were prepared by hydrothermal treatment at 180°C for 24 and 14 h, respectively. Titanate isopropoxide or tetrabutyl titanate was used as the precursor for (001)-faceted TiO$_2$ nanocrystals with hydrofluoric acid (HF) as the facet controlling agent, and titanium tetrachloride for (101)-faceted TiO$_2$ nanocrystals with hydrochloric acid (HCl) as the crystallographic controlling agent. The hydrogenation of TiO$_2$ nanocrystals was performed at 450°C with an initial hydrogen pressure of 7.0 MPa, and hydrogen storage capacities of 1.0 wt.% and 1.4 wt.% were achieved for (001)- and (101)-faceted TiO$_2$ nanocrystals, respectively.[18] The hydrogenated (001)- and (101)-faceted TiO$_2$ nanocrystals had different colors. The former was blue, and the latter showed black color. In addition, it was found that on combining calculations data (i) hydrogen incorporation through (101) facets was more favorable, and (ii) hydrogen stayed in the interstitial sites between titanium oxygen octahedrons.[18]

Lu et al. studied the hydrogenation of commercial P25 TiO_2 for different reaction time.[19] Typically, 0.5 g of P25 TiO_2 powders were used as the pristine materials, and kept in a vacuum for 24 h and then hydrogenated in 35 bar hydrogen atmosphere at room temperature. The color of the white P25 turned into pale yellow, gray and black after hydrogenation for 3, 15 and 20 days. Heavy surface disorder was observed with an obvious core–shell structure.[19] However, a slight phase change was found in the hydrogenated P25 as the hydrogenation time was elongated to ≥ 15 days.[19]

Leshuk et al. also investigated the parameters affecting the hydrogenation reaction in detail.[20,21] Synthesis protocols, hydrogenation temperature and pressure were revealed to be critical parameters in controlling the structure, coloration and photocatalytic activity of TiO_2 nanoparticles.[20,21] H_2 likely reacted with pre-existent reactive sites in TiO_2 crystals, possibly dangling bonds or other crystalline disorder, wherein some samples became black as a result of hydrogenation, while others turned into different colors or even remained unchanged in appearance.[21] Hydrogenation pressure and temperature were disclosed to be fundamentally important parameters in controlling the coloration of hydrogenated TiO_2 nanomaterials.[21]

Ambient- and low-pressure hydrogenation: The hydrogenated TiO_2 can be obtained in a low-pressure H_2 atmosphere, but commonly under relative high temperature (≥400°C). Interestingly, phase transformation from anatase to rutile can be observed at temperatures lower than the usual phase change temperature during the hydrogenation. For example, a recent research reported on black hydrogenated TiO_2 nanoparticles with unique crystalline core/disordered shell morphology also found the phase transformation at 500°C. In a typical synthesis, ~0.3 g of the commercial amorphous TiO_2 was initially treated under vacuum (10^{-5} mbar) and then heated at 200°C with flowing O_2 for 1 h to remove the possible molecular species adsorbed onto the surface of the TiO_2 nanoparticles; the hydrogen reduction was performed in H_2 flow at 500°C for 1 h.[22] Fast cooling in inert environment until room temperature resulted in black TiO_2

nanoparticles, while very slow cooling rate or instantaneous exposure to air resulted in a gray coloration. Hydrogen reduction triggered the phase change below the usual phase change temperature. Black TiO_2 nanoparticles obtained at 500°C had 81% anatase and 19% rutile phases, whereas white TiO_2 prepared in flowing O_2 at 500°C for 1 h was 100% anatase. However, reduction of TiO_2 at 400 and 450°C generated samples that were 100% anatase.[22] The interaction between TiO_2 host matrix and hot H_2 molecule gave rise to oxygen vacancies that overcame the activation energy of TiO_2 lattice rearrangement and accelerated the phase transformation from anatase to rutile.[22,23]

Hydrogenated black rutile TiO_2 nanowire arrays were synthesized in a hydrogen atmosphere (ambient pressure) at various temperatures in a range of 200–550°C for 30 min.[24] Rutile TiO_2 nanowire arrays were grown on a fluorine-doped tin oxide (FTO) glass substrate using hydrothermal method at 150°C for 5 h, followed by being annealed in air at 550°C for 3 h.[24] Figure 2(a) shows the scanning electron microscope (SEM) image of the vertically aligned TiO_2 nanowire arrays on FTO substrate. The color of the hydrogenated rutile TiO_2 nanowires turned yellowish green at 350°C and black at 450°C and above (Fig. 2(b)).

Ordered mesoporous black TiO_2 was prepared through an evaporation-induced self-assembly method combined with an ethylenediamine

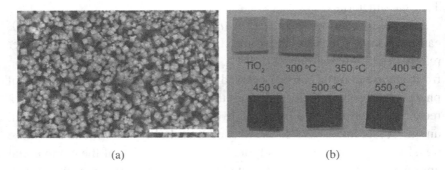

(a)   (b)

Fig. 2. (a) SEM image of vertically aligned TiO_2 nanowire arrays prepared on a FTO substrate. Digital pictures of pristine and hydrogenated TiO_2 nanowires on FTO substrate annealed in a hydrogen atmosphere at various temperatures (300°C, 350°C, 400°C, 450°C, 500°C and 550°C).[24] Reprinted with permission from Ref 24. Copyright 24, American Chemical Society.

encircling process, followed by hydrogenation.[25] Ordered mesoporous TiO_2 turned a black color after hydrogenation at 500°C for 3 h under H_2 flow of normal pressure, whereas only gray TiO_2 was obtained after hydrogenation at the same conditions if the porous TiO_2 was synthesized by normal process.[25] The surface functionalities and surface structural defects were very important in determining the colorization of the hydrogenated TiO_2 materials.

2.1.2 Hydrogenation in hydrogen–argon atmosphere

Liu et al. reported a series of hydrogenated TiO_2 nanotube arrays under various conditions: (i) pristine anatase TiO_2 nanotube arrays annealed in air at 450°C for 1 h; (ii) TiO_2 nanotube arrays treated in Ar or H_2/Ar at 500°C for 1 h under atmospheric pressure; (iii) high-pressure hydrogenated TiO_2 nanotube arrays in pure hydrogen (20 bar, 500°C for 1 h); and (iv) high-pressure but mild-temperature hydrogenated TiO_2 nanotube arrays (pure H_2, 20 bar, 200°C for five days).[26] The color of the TiO_2 nanotubes turned from white into black in H_2/Ar, deep purple in Ar, light blue in high-pressure H_2 at 500°C, and gray in high-pressure H_2 at 200°C. The pristine TiO_2 nanotubes were obtained by an anodic oxidation method. To obtain TiO_2 nanotubes, titanium foils were anodized at 60 V for 15 min in a two-electrode configuration with Pt as the counter electrode and ethylene glycol with addition of H_2O (1 M) and NH_4F (0.1 M) as the electrolyte.[26]

Lu et al. prepared a series of hydrogenated TiO_2 nanomaterials with different shapes (anatase TiO_2 nanotubes, anatase TiO_2 nanoparticles and rutile TiO_2 nanorods).[27] Anatase TiO_2 nanotubes were prepared by an anodic oxidation method. Briefly, titanium foils were anodized at 60 V for 0.5 h in the electrolyte of ethylene glycol containing 0.5 wt.% ammonium fluoride and 4–6 vol% deionized water with Pt as the counter electrode.[27] The hydrogenation was carried out by annealing various TiO_2 nanomaterials at 450°C for 1 h in a 5% H_2/95% Ar atmosphere with a flow rate of 100 sccm.[27] The rutile TiO_2 nanorods were synthesized by heating titanium n-butoxide (0.5 mL) in HCl (18.25%) solution at 150°C for 5 h, followed by annealing in air at 550°C for 3 h.[27] The anatase TiO_2 nanoparticles were bought from Aldrich with grain size of 50 nm.[27] All the hydrogenated TiO_2 nanomaterials turned black (Fig. 3).

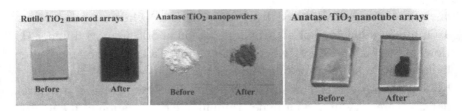

Fig. 3. Digital images of rutile nanorods, anatase TiO_2 nanoparticles and nanotubes before and after hydrogenation at 450°C for 1 h under a gas flow of 5% H_2/95% Ar.[27] Reprinted with permission from Ref. 27. Copyright 2012, Wiley-VCH.

Danon et al. treated TiO_2 nanotubes under a flow of 5% H_2 in Ar at atmospheric pressure at 350°C for 3 h. Black powders were obtained in a stainless reactor, whereas only blue powders were synthesized in a quartz reactor.[28] The pristine TiO_2 nanotubes were prepared by a hydrothermal method in 10 M NaOH solution at 120°C for 48 h using anatase TiO_2 powders as the raw materials.[28]

2.1.3 *Hydrogenation in hydrogen–nitrogen atmosphere*

Zhu et al. prepared black TiO_2 through Pt-enhanced hydrogen reduction with Pt nanoparticles loaded on the P25 TiO_2 nanoparticles.[29] Pt/P25 was first prepared via conventional impregnation followed by reduction with sodium borohydride.[29] Hydrogenation of Pt/TiO_2 was carried out in a 8% H_2/N_2 atmosphere at 200–700°C for 4 h.[29] The hydrogenation reduction started from 160°C, and became apparent at 400, 500, 700 and 750°C due to the hydrogen spillover effect from Pt to TiO_2.[29]

2.2 Hydrogen plasma treatment

Wang et al. synthesized H-doped black titania with a core–shell structure (TiO_2@$TiO_{2-x}H_x$).[30] Black $TiO_{2-x}H_x$ was prepared from P25 by hydrogen plasma at 500°C for 4–8 h with a plasma input power of 200 W.[30] The color of the pristine TiO_2 turned from white to black as it was converted into $TiO_{2-x}H_x$ during the hydrogen plasma treatment.[30] In addition, Wang et al. also synthesized black TiO_{2-x} by annealing amorphous TiO_2 under an oxygen flow at 200°C, followed by hydrogenation at 500°C under a hydrogen flow.[30] Teng et al. prepared black TiO_2 nanotubes by hydrogen plasma assisted chemical vapor deposition.[31] The pristine TiO_2 nanotubes were prepared by a hydrothermal method in 10 M NaOH solution at 120°C for 12 h using Degussa P25 nanoparticles as the raw materials, then hydrogenation was performed in a hot-filament chemical vapor deposition apparatus at 350 and 500°C for 3 h with hydrogen as reaction gas.[31] Yan et al. also prepared black hydrogenated anatase TiO_2 nanoparticles using hydrogen plasma treatment.[31] The hydrogen plasma treatment was performed on commercial anatase TiO_2 nanoparticles at 390°C for 3 h, with the inductively coupled plasma power of 3,000 W, the chamber pressure of 26.5–28.3 mTorr, and the H_2 flow rate of 50 sccm.[31]

The color and properties of the hydrogenated P25 TiO_2 nanoparticles depended on the retention time during the plasma treatment.[32] The hydrogen plasma treatment was performed at 150°C with the inductively coupled plasma power of 3,000 W, the chamber pressure of 25.8–27.1 mTorr, and the H_2 flow rate of 50 sccm.[32] As shown in Fig. 4, the white P25 TiO_2 nanoparticles turned gray within 3 min, and became black after 20 min. Amorphous shell was observed by HRTEM, and the thickness of the amorphous shell increased with increasing the retention time.

Fig. 4. Digital and HRTEM images of (a) the pristine TiO_2 and (b–f) the hydrogenated TiO_2 prepared by hydrogen plasma treatment after different times of (b) 0.5, (c) 1, (d) 3, (e) 5 and (f) 20 min. Based on the retention time, the samples are denoted as H–TiO_2-0.5, H–TiO_2-1, H–TiO_2-3, H–TiO_2-5 and H–TiO_2-20.[32] Reprinted with permission from Ref. 32. Copyright 2014, the Royal Society of Chemistry.

2.3 Chemical reduction

2.3.1 *Aluminum reduction*

Huang's group reported on the synthesis of series of black TiO_2 nanomaterials by aluminum reduction.[33–38] In a typical synthesis, aluminum and pristine TiO_2 nanoparticles were placed separately in a two-zone tube furnace (Fig. 5(a)), the pressure in the tube was controlled at a base pressure below 0.5 Pa, and then aluminum was heated at 800°C while TiO_2 was heated at 300–500°C for 6 h.[34] As shown in Fig. 5(b), black TiO_2 nanoparticles were produced on a large scale using aluminum reduction method. The pristine TiO_2 nanocrystals were highly crystallized (Fig. 5(c)).[34] A unique crystalline/amorphous core–shell structure was observed on all aluminum-reduced TiO_2, and the thickness of the disordered outer layer increased with the aluminum-reduction temperature (Fig. 5(d)–5(f)).[34]

Fig. 5. (a) Schematic low-temperature reduction of TiO_2 in a two-zone furnace, (b) digital images of white and black titania and (c–f) HRTEM images of TiO_2 nanocrystals before (c) and after (d–f) the Al reduction at different temperatures for 6 h.[34] Reprinted with permission from Ref. 34. Copyright 2013, the Royal Society of Chemistry.

2.3.2 NaBH$_4$ reduction

Kang et al. synthesized black TiO$_2$ nanotubes by chemical reduction using NaBH$_4$ as the reduction agent.[39] The starting TiO$_2$ nanotubes were obtained by anodizing a Ti foil in an ethylene glycol solution containing 0.3 wt.% NH$_4$F and 2 vol.% H$_2$O under 80 V for 30 min at 5°C with a graphite cathode, followed by annealing at 450°C for 3 h.[39] Black TiO$_2$ nanotubes were obtained after the reduction reaction performed in 0.1 M NaBH$_4$ solution for 10–60 min at room temperature.[39]

2.4 Chemical oxidation

Grabstanowicz et al. reported black rutile TiO$_2$ powder that was prepared by oxidizing TiH$_2$ in H$_2$O$_2$ followed by calcinations in Ar gas.[40] Typically, 15 mL of H$_2$O$_2$ was added to 10 mL of aqueous suspension containing 0.96 g of TiH$_2$ powder and stirred for 3 h at room temperature to obtain a miscible gel-like slurry; another 12 mL of H$_2$O$_2$ was added to the slurry and stirred for 4 h, followed by adding another 15 mL of H$_2$O$_2$ and stirring for another 16 h to form a yellow gel; the gel was vacuum-desiccated overnight, then placed in an oven at 100°C for 12–20 h, and finally calcinated at 630°C for 3 h in Ar.[40]

Xin and co-authors prepared colored anatase TiO$_2$ via a solvothermal process using TiH$_2$ and H$_2$O$_2$ as Ti source and oxidant, respectively, followed by post-annealing at different temperatures.[41] Typically, a yellowish gel was first obtained by reacting TiH$_2$ and H$_2$O$_2$ for 12 h; then the gel was diluted using ethanol, the pH of the mixture was adjusted to 9.0 by NaOH, and NaBH$_4$ as an antioxidant was added to the resulting mixture; after the solvothermal treatment at 180°C for 24 h, the collected sample was washed by HCl, water and ethanol; light blue TiO$_2$ nanocrystals were obtained after the precipitate was dried in vacuum for 12 h; finally, post-annealing treatment was carried out at 300–700 °C for 3 h under nitrogen flow.[41] The annealing temperature was found to be a crucial factor affecting the color of the as-prepared TiO$_2$ nanocrystals.[41] As shown in Fig. 6, light brown, brown, black, dark brown and shallow dark brown TiO$_2$ nanocrystals were obtained at 300°C, 400°C, 500°C, 600°C and 700°C, respectively.[41] No phase change was observed, whereas obvious surface

Fig. 6. Digital images of Ti^{3+} self-doped TiO_2 samples prepared with post-annealing treatment in the range of 300–700°C for 3 h in a N_2 gas flow.[41] Reprinted with permission from Ref. 41. Copyright 2015, Elsevier.

disorder or surface amorphous layer was detected on each single TiO_2 nanocrystal.[41]

2.5 Electrochemical reduction

Black TiO_2 nanotubes were synthesized through electrochemical reduction of anatase TiO_2 nanotubes in ethylene glycol electrolytes.[42–45] In a typical synthesis, anatase TiO_2 nanotubes were firstly synthesized by anodization in an aged ethylene glycol solution of 0.2 M HF and 0.12 M H_2O_2 at a constant voltage of 80 V for 2 h with Pt gauze as the counter electrode and Ti foil as the working anode, then annealed in air at 450°C in air for 4 h, and finally underwent electrochemical reduction at −40 V for 200 s in an ethylene glycol solution containing 0.27 wt.% NH_4F after a brief activation treatment in the aforementioned anodization electrolyte (typically, at 60 V for 30 s, or 4 V for 600 s) to obtain black TiO_2 nanotubes. TiO_2 nanotubes were first prepared with a under a constant voltage or current (typically, 80 V for 7200 s, or 4 mA for 5,000 s) in an "aged" ethylene glycol solution under 60 V for 12 h with 0.2 M HF and 0.12 M H_2O_2.[42] The resulting black TiO_2 was highly stable with a significantly narrower bandgap and higher electrical conductivity.[42]

Xu et al. prepared electrochemically hydrogenated black TiO_2 nanotubes.[45] The pristine anodic TiO_2 nanotubes were prepared at 150 V for 1 h in an ethylene glycol electrolyte containing 0.3 wt.% NH_4F and 10 vol.% H_2O with carbon rod as the cathode and Ti foil as the anode.[45] After the products were crystallized in ambient air at 150°C for 3 h, and then 450°C for 5 h, the electrochemical reduction was performed at 5 V for 5–40 s in 0.5 M Na_2SO_4 aqueous solution at room temperature with the TiO_2 nanotubes as the cathode and Pt as the anode.[45]

2.6 Other methods

Myung et al. prepared black TiO_2 nanoparticles by annealing a yellow TiO_2 gel in Ar gas at 400–600°C for 5 h.[46] A yellow TiO_2 gel was obtained by adding high-purity $TiCl_4$ dropwise to a water–ethanol solution containing HF (pH < 2) and urea (urea was added to delay the hydrolysis of Ti^{4+} to $Ti(OH)_4$.) at 0°C with continuous stirring for 4 h, and then evaporating the solution slowly at 80°C.[46] The color of the annealed TiO_2 nanoparticles was highly dependent on the annealing temperature and the concentration of HF in the initial reaction solution. Dark yellow, gray and black TiO_2 nanoparticles were obtained at 200°C, 300°C, and ≥ 400°C, respectively, and the color turns black on increasing the concentration of HF.[46] The crystallinity of TiO_2 nanoparticles was enhanced on increasing the annealing temperature, and no phase change was observed.

Dong et al. prepared black TiO_2 nanotubes were directly derived from anodic TiO_2 nanotubes through annealing in air.[47] TiO_2 nanotubes were prepared by two-step anodization on Ti foil at 60 V for 10 h in ethylene glycol containing 0.25 wt.% NH_4F and 2 vol.% distilled water.[47] The anodized Ti foil was cleaned in ethanol and distilled water, dried at 150°C, and then sintered at 450°C for 1 h in ambient atmosphere.[47] A layer of black TiO_2 was obtained on the substrate after removing the top oxide layer.[47]

Ionothermal synthesis of black single-crystal anatase TiO_2 was reported by Li et al.[48] In a typical synthesis, 12 mL of N,N-dimethylformamide, 0.6 g of $LiAc·2H_2O$ and 18 mL of glacial HAc were mixed in a 100 mL pyrex beaker and magnetically stirred for 15 min to get a clear solution, the mixture was then transferred to an autoclave (50 mL) to which a piece of Ti foil and 2 mL ionic liquid (1-methyl–imidazolium tetrafluoroborate) were added; finally, the mixture was kept at 200°C for 24 h.[48] Black TiO_2 was obtained with a surface thin layer on each single nanocrystal.

3. Properties of Black TiO_2 Nanomaterials

Black TiO_2 nanomaterials have been obtained with various methods, such as hydrogen thermal treatment, hydrogen plasma treatment, chemical reduction, chemical oxidation, electrochemical reduction, etc. Although those black TiO_2 nanomaterials have similar appearance, their structures may differ significantly owing to the different preparation methods and different reaction parameters. Therefore, black TiO_2 nanomaterials from different research groups worldwide possess different chemical and physical properties. Some of these structural properties, if not all, have been used to explain the color change from white to black.

3.1 Structural disorder near the surface

Crystalline/disordered core/shell structures were frequently observed in black TiO_2 nanoparticles,[13–16,19,22,30,34,41] though sometimes the surface disorder was not found.[27] HRTEM were often used to differentiate the disordered phase from the crystalline phase in the black TiO_2 nanoparticles.[13–16,19,22,30,34,41] As shown in Fig. 7, HRTEM and line analyses were employed to comparatively study the structures of the pristine white TiO_2 nanoparticle and the hydrogenated black TiO_2 nanoparticle.[15] The pristine white TiO_2 nanoparticle displayed clearly-resolved and well-defined lattice fringes throughout the whole nanocrystal, and the distance between the adjacent lattice planes was typical for anatase (0.352 nm).[15] In sharp contrast, the black TiO_2 nanoparticle had a crystalline-disordered core–shell structure,[15] and a structural deviation from the standard crystalline anatase was readily seen at the outer layer, where the straight lattice line was bent at the edge of the nanoparticle and the plane distance was no longer uniform.[15] Electron diffraction (ED) was another method to identify the amorphous structure from the crystalline phase.[49] The nanoparticles containing only crystalline phase gave clear diffraction patterns made of diffraction dots or rings, but instead the nanoparticles having both crystalline and disordered phases produced cloudy diffraction patterns.[49] Fast Fourier transform (FFT) from the HRTEM images was also a useful technique to distinguish the amorphous phase from the crystalline phase in black TiO_2 nanoparticles.[22]

X-ray diffraction (XRD) was used to analyze the crystalline phase structures in black TiO_2 nanoparticles.[13,14,19] The XRD peaks of black TiO_2 nanoparticles shifted to higher diffraction angles compared to pristine white TiO_2 nanoparticles, suggesting that the reduction of the interplanar

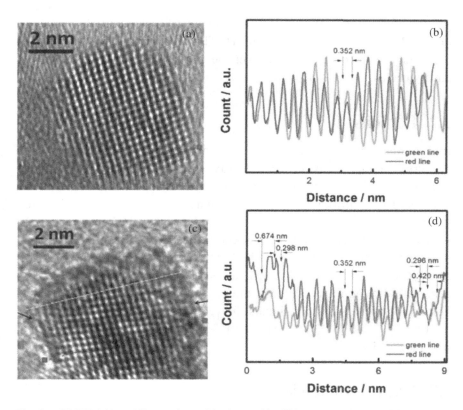

Fig. 7. HRTEM (a) and line analyses (b) of one white TiO_2 nanoparticle, HRTEM (c) and line analyses (d) of one black TiO_2 nanoparticle. The zeros of the axis in (b) and (d) correspond to the left ends of the lines in **a** and c. The red and green curves in b and d correspond to the red and green lines in a and c.[15] Reprinted with permission from Ref 15. Copyright 2013, Nature Publishing Group.

distances of the crystalline phase.[14] However, XRD is a bulk analysis technique and only sensitive to the crystalline phase. Raman spectroscopy was used as a complimentary structural characterization technique in probing the amorphous or disordered structures in black TiO_2 nanoparticles.[13] In Raman spectra of black TiO_2 nanoparticles, the scattering peaks became much weaker in intensity,[13,27] broader in width,[13,19,20,22,27,34] shifted to higher wavenumbers,[18,20,22,33,34] and additional small vibrational modes were observed.[13] The combination of XRD and Raman results provided integrated structural information of the black TiO_2 nanoparticles. In addition, XRD was also combined with HRTEM to estimate the volume and percentage of disordered phase over the crystalline phase in black TiO_2 nanoparticles.[14]

3.2 Ti^{3+} ions

In a number of reported black TiO_2 nanomaterials, Ti^{3+} ions were not detected using conventional X-ray photoelectron spectroscope (XPS).[13,15,22,24,27] However, many papers argued that Ti^{3+} ions were present and could be confirmed in black TiO_2 nanomaterials by XPS measurement.[33,38,41,43,46,48] More commonly, Ti^{3+} ions in black TiO_2 nanomaterials were detected by electron paramagnetic resonance (EPR) or electron spin resonance (ESR) spectroscope.[26,30,32,35,36,40,41,48] For example, the presence of EPR signal at about g = 1.957, a characteristic of paramagnetic Ti^{3+} centers, indicated the presence of the Ti^{3+} spins in the TiO_{2-x}, whereas the absence of the similar ERP signals showed the absence of Ti^{3+} spins in the pristine TiO_2 and $TiO_{2-x}H_x$.[30] However, in some cases, even EPR did not find the Ti^{3+} ions.[47,50]

3.3 Oxygen vacancies

Oxygen vacancies were suggested to accompany with black TiO_2 nanomaterials. Again, ESR or EPR was widely used to detect the oxygen vacancies.[26,29,31,35,37,39] For example, the ESR signal at g ≈ 2.02 was usually attributed to the oxygen vacancies.[26,29,31,35,37,39] Raman spectroscopy was also used to identify the presence of the oxygen vacancies,[23,33,34,38,46–48] and sometimes XRD analysis was also used.[22,34,48]

3.4. Ti–OH groups

XPS was used to reveal the change in –OH groups of hydrogenated TiO_2 nanomaterials. A shoulder peak of Ti–OH in the O 1s XPS spectrum indicated the increase of –OH groups in the hydrogenated black TiO_2.[13,24] An example is given in Fig. 8(a). Fourier transform infrared (FTIR) spectroscopy was employed to study the change in surface –OH groups by

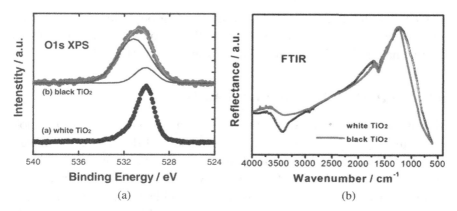

Fig. 8. (a) O 1s XPS spectra of the white and black TiO$_2$ nanocrystals. The red and black circles are XPS data. The green curve is the fitting of experimental data for black TiO$_2$ nanocrystals, which can be decomposed into a superposition of two peaks shown as blue curves. Reproduced from Ref. 13. Copyright The American Association for the Advancement of Science, 2011. (b) FTIR reflectance spectra of hydrogenated black TiO$_2$ nanocrystals compared with white TiO$_2$ nanocrystals. Reproduced from Ref. 15. Copyright 2013, The American Association for the Advancement of Science.

comparing the magnitude of the intensity of the peak corresponding to the –OH vibrational band.[15,27,30,49] One example is shown in Fig. 8(b). Black TiO$_2$ nanomaterials from hydrogen plasma treatment displayed extra peaks at 3685, 3670, 3645 cm^{-1} and at 3710 cm^{-1}.[30] However, there were also some exceptions. The –OH response in O 1s XPS spectrum did not show any change before and after hydrogenation in the study by Liu et al.[27] and the –OH signals in the O 1s XPS spectrum and in the FTIR spectrum even showed a decrease in a few cases.[15,27,49] In addition, ^{1}H nuclear magnetic resonance (NMR) technique was also used to analyze the surface –OH groups.[15,30,49]

3.5. Ti–H groups

Zheng et al. found that hydrogenated TiO$_2$ nanowire microspheres exhibited one shoulder peak at the lower binding energy side of the broader Ti 2p peak and attributed it to the surface Ti–H bonds formed under hydrogen atmosphere.[51] Formation of surface Ti–H bonds was at the expense of surface Ti–OH and thus reduced the OH groups in TiO_2.[51] Wang et al. attributed the peak at 457.1 eV in the Ti $2p$ XPS spectrum of the hydrogenated black TiO_2 nanocrystals to surface Ti–H bonds.[30]

3.6 Change in the valence band

Shift in the valence band was sometimes observed in the black TiO_2 nanomaterials.[13,22,27,33,35,36,39,42] For example, a red shift of the valence band was observed from valence-band XPS spectrum for the hydrogenated TiO_2 nanoparticles treated at 200°C for four days under 20 bar H_2[13] or heated at 500°C for 1 h under hydrogen atmosphere,[22] in black brookite TiO_2 nanoparticles,[35] black rutile TiO_2 nanoparticles,[36] and black TiO_2 nanotubes[33] prepared with the Al reduction method, in the black anatase TiO_2 nanotubes by $NaBH_4$ reduction,[39] in the hydrogenated TiO_2 nanotubes treated at 450°C for 1 h in a reducing 5% H_2 and 95% Ar[27] and in the black TiO_2 nanotubes electrochemically hydrogenated.[42] Ti^{3+} ions was found to not contribute to the extra bandgap states of the hydrogenated TiO_2 nanocrystals, as the these midgap states disappeared on the appearance of Ti^{3+}.[15]

3.7 Theoretical consideration of hydrogenated black TiO_2 nanomaterials

Theoretical calculations were largely helpful to clarify the structural properties and enhanced performances. Chen et al. suggested that the H atoms were bonded to O and Ti atoms, and two midgap states at about 3.0 eV and 1.8 eV, respectively, were predicted, with the higher-energy midgap made of Ti $3d$ orbitals, and the lower-energy midgap composed of O $2p$ and Ti $3d$ orbitals.[13] The hydrogen $1s$ orbital mainly passivated the dangling bonds and stabilized the lattice disorders, which contributed to the midgap states.[13]

Lu et al. investigated the hydrogenation effect on the structure and photocatalytic properties of anatase TiO_2 (101) and (001) surfaces using Density Functional Theory–Perdew-Burke-Ernzerhof (DFT–PBE) calculations.[52] They proposed that the high photoactivity of disorder-engineered TiO_2 nanocrystals was ascribed to surface effects.[52] They found that hydrogen atoms were chemically absorbed on both Ti_{5c} and O_{2c} atoms for

(101), (001) and (100) surfaces by taking into account the synergistic effect of Ti–H and O–H bonds.[52] The hydrogenation-induced lattice distortions on (101) and (100) surfaces of nanoparticles enhanced the intraband coupling within the valence band, while the (001) surface was not largely affected.[52] The atoms not only induced the lattice disorders but also interacted strongly with the O $2p$ and Ti $3d$ states, resulting in considerable contribution to the midgap states.[52] The optical absorption was dramatically red shifted due to the midgap states and the photogenerated electron–hole separation was substantially promoted as a result of electron–hole flow between different facets of hydrogenated nanoparticles.[52]

Liu and co-authors later pointed out that hydrogen passivation of Ti and O dangling bonds on the surface of hydrogenated black TiO_2 contributed to the shifts of the conduction band minimum (CBM) and the valence band maximum (VBM), even surface disorder.[16] The CBM did not change upon the O-sublattice distortion but underwent red shift upon the Ti-sublattice distortion; the VBM blue shifted in either case.[16] The lattice disorder was mainly due to the O distortion.[16] Hydrogenation played an important role in reducing the distortion energy in raising the VBM.[13,16] They examined further the hydrogenation effect by comparing two bulk anatase models, either with an interstitial H_2 molecule or with two H atoms bonded to O and Ti separately.[16] The CBM electrons were not sensitive to local lattice distortions nor to H bonding induced changes.[16] The midgap states changed upon both lattice distortion and H bonding.[16] The lattice distortion caused by H in bulk anatase was not very stable and might release hydrogen from the lattice even at room temperature.[16]

Deng *et al.* studied the effect of hydrogen passivation on the electronic structure of ionic semiconductor nanostructures.[53] First-principles band structure theory calculation and systematic comparison between PH-passivated and real-hydrogen–passivated (RH-passivated) semiconductor surfaces and nanocrystals showed that unlike PH passivation that always increases the bandgap with respect to the bulk value, RH passivation of the nanostructured semiconductors can either increase or decrease the bandgap, depending on the ionicity of the nanocompounds.[53] For anatase TiO_2, occupied gap states from the Ti–H bond with Hs and Tis and d characters were created at ~0.7 eV above the host VBM and the effective bandgap was reduced.[53] The position of occupied H-induced state (bonding cation-H state) near the VBM state was not obviously affected by the anion.[53] The first-principle calculations by Liu et al. suggested that the absolute value of the O_2 adsorption energy was increased by H atoms on the hydrogenated anatase (101) surface or at subsurface sites, and the dissociation barriers of O_2 on an anatase surface were decreased with two H atoms at the subsurface sites or with an H atom on the surface and a subsurface H atom.[54]

4. Applications of Black TiO_2 Nanomaterials
4.1 Photocatalysis

Chen et al. found that the hydrogenated black TiO_2 nanoparticles had much higher photocatalytic activities in decomposing organic pollutants (methylene blue and phenol) and generating H_2 from water/methanol solution.[13] Wang and co-authors reported that hydrogenated black TiO_2 nanowires displayed enhanced photoelectrochemical water-splitting performance.[24] Zheng et al. found that hydrogenated brown anatase nanowire microspheres showed highly enhanced photocatalytic activity in photodecomposition of 2,4-dichlorophenol in the visible-light region.[51] Wang and co-authors produced black TiO_2 under hydrogen plasma demonstrated large improvement in photocatalytic decomposition of methyl-orange.[30] Lu et al. showed hydrogenated TiO_2 nanocrystals displayed improved photocatalytic activity in hydrogen generation.[19]

Cui et al. demonstrated that black anatase TiO_2 nanotubes from Al reduction had improved photoelectrochemical water-splitting performance.[33] Wang et al. found that black anatase nanoparticles from Al reduction showed greatly enhanced photocatalytic activity in decomposing methyl orange under both UV and visible-light irradiations, showing high mineralization efficiency.[34] Zhu et al. showed that black brookite nanoparticles from Al reduction had enhanced photocatalytic activities in decomposing methyl orange and methylene blue under sunlight.[35] Yang et al. showed the black rutile TiO_2 nanoparticles from Al reduction with improved photoelectrochemical water-splitting performance under both UV and visible-light irradiations.[36] Kang et al. reported that black anatase TiO_2 nanotubes from $NaBH_4$ reduction showed enhanced photoelectrochemical performance.[39] Black TiO_2 nanoparticles from the oxidation of TiH_2 by H_2O_2 had enhanced photocatalytic activities in both methylene blue decomposition and hydrogen generation under visible-light irradiation.[40] Overall, the enhanced photocatalytic and photoelectrochemical activities of black TiO_2 nanomaterials have been frequently attributed to the existence of oxygen vacancies,[24,34,40,35–37] Ti^{3+} ions,[33–37] surface Ti–OH groups,[13,24] surface Ti–H groups,[13,51] narrower bandgap,[51] better charge separation[33–37] or the combination effect of these species.[33–37]

4.2 Lithium-ion batteries

Lu *et al.* reported on the hydrogenated black TiO_2 nanotube arrays as high-rate anodes for lithium-ion microbatteries.[27] The hydrogenated black TiO_2 nanotube arrays showed a higher capacity, a better rate capability, and a better cycling stability compared to pristine white TiO_2 nanotube arrays.[27] Yan and co-authors reported on the fast lithium storage performance of hydrogenated anatase TiO_2 nanoparticles (H–TiO_2) prepared by a hydrogen plasma treatment.[56] The scan-rate dependence of the cyclic voltammetry analysis revealed that the improved rate capability of H–TiO_2 resulted from the enhanced pseudocapacitive lithium storage on the disordered surface layers.[55] In addition, the introduction of Ti^{3+} species (oxygen vacancies) improved the electric conductivity of H–TiO_2, which also contributed to the fast lithiation/delithiation process.[55]

Myung *et al.* found that the presence of Ti^{3+} defects in black anatase TiO_2 narrowed its bandgap energy to a semiconductor level of 1.8 eV, resulting in high electrical conductivity of 8×10^{-2} S cm^{-1}.[46] The high electrical conductivity ensured an ultrafast Li^+ insertion into and extraction from the host structure of anatase TiO_2, delivering a discharge capacity of 127 mAh g^{-1} at 100°C (20 A g^{-1}) with approximately 86% retention after 100 cycles at 25°C.[46]

4.3 Supercapacitors

Li *et al.* reported on the use of anodic black TiO_2 nanotubes as supercapacitor electrode.[42] A 42-fold increase in specific capacitance was observed.[42] The remarkable enhancement can be attributed to their high conductivities and an increased density of hydroxyl groups on the surface of TiO_2.[42] Electrochemically self-doped TiO_2 nanotube arrays reported by Zhou et al. also exhibited excellent capacitive properties.[43] The capacitance of the self-doped TiO_2 nanotube arrays was 39 times higher than that of the pristine TiO_2 nanotube arrays, and the self-doped TiO_2 nanotube arrays exhibited excellent long-term stability with capacitance retention of 93.1% after 2,000 cycles.[43]

5. Summary and Prospective

Black TiO_2 has triggered worldwide attention since its discovery in 2011. So far, a variety of preparation methods, e.g. hydrogenation, hydrogen plasma treatment, chemical reduction, chemical oxidation, and electrochemical reduction, have been explored to achieve black TiO_2 nanomaterials with different sizes, shapes and textures. The surface disorder caused by the surface structural rearrangement and surface chemical changes in Ti^{4+} and O^{2-} ions extends the optical absorption of black TiO_2 nanomaterials to visible-light spectrum, along with bandgap narrowing and improvement in electrical conductivity. These superior properties have attracted tremendous interest in their applications in photocatalysis, lithium-ion batteries and supercapacitors. In addition, applications of surface-disordered TiO_2 nanomaterials have been expanded to the fields of fuel cells, photoelectrochemical sensors, field emission electrodes and microwave absorbers.

Despite those great progresses in preparation and applications of black TiO_2 nanomaterials, there are quite a few challenges that black TiO_2 nanomaterials are facing. For example, different black TiO_2 nanomaterials show differences in photocatalytic activity. Even small changes in reaction parameters can result in considerable differences in optical and electronic properties of black TiO_2 nanomaterials. This suggests that the structure and properties of pristine TiO_2 nanomaterials play key roles in determining the properties of the black TiO_2 nanomaterials. Therefore, it is still challenging but of high importance to understand the structural evolution of black TiO_2 nanomaterials during the reaction processes, along with disclosing the effect of physical/chemical properties of the pristine TiO_2

on the properties of the corresponding black TiO_2. Another issue that needs to be addressed is the identification of the surface chemical compositions of the black TiO_2 nanomaterials. To date, it is still difficult to clearly identify the presence of Ti^{3+} ions and Ti–H groups. New technologies need to be applied to resolve these challenges. In addition, the relationship between defects and physicochemical properties of black TiO_2 nanomaterials needs to be precisely related. It is still unclear whether one specific type of defect favors one or several specific applications, or it is the synergistic effect of different defects that result in the highly improved performance of the black TiO_2 nanomaterials in many applications. This chapter therefore, aims to call for more efforts to advance the development of black TiO_2 nanomaterials.

Acknowledgments

Xiaobo Chen thanks the support from the College of Arts and Sciences, the University of Missouri–Kansas City and the University of Missouri Research Board. Lihong Tian thanks the National Natural Science Foundation of China (No. 51302072) and China Scholarship Council for their financial supports.

References

1. Y. Ma, X. Wang, Y. Jia, X. Chen, H. Han and C. Li, *Chem. Rev.* **114**, 2014, 9987.
2. M. Pelaez, N. T. Nolan, S. C. Pillai, M. K. Seery, P. Falaras, A. G. Kontos, P. S. M. Dunlop, J. W. J. Hamilton, J. A. Byrne, K. O'Shea, M. H. Entezari and D. D. Dionysiou, *Appl. Catal. B-Environ.* **125**, 2012, 331.
3. H. Chen, C. E. Nanayakkara and V. H. Grassian, *Chem. Rev.* **112**, 2012, 5919
4. Y. Bai, I. Mora-Seró, F. D. Angelis, J. Bisquert and P. Wang, *Chem. Rev.* **114**, 2014, 10095.
5. Z. F. Yin, L. Wu, H. G. Yang and Y. H. Su, *Phys. Chem. Chem. Phys.* **15**, 2013, 4844.
6. X. Chen and S. S. Mao, *Chem. Rev.* **107**, 2007, 2891.
7. X. Chen, C. Li, M. Grätzel, R. Kostecki and S. S. Mao, *Chem. Soc. Rev.* **41**, 2012, 7909.
8. X. Chen and C. Burda, *J. Am. Chem. Soc.* **130**, 2008, 5018.

9. S. Sandoval, J. Yang, J. G. Alfaro, A. Liberman, M. Makale, C. E. Chiang, I. K. Schuller, A. C. Kummel and W. C. Trogler, *Chem. Mater.* **24**, 2012, 4222.
10. J.-M. Wu and M.-L. Tang, *Nanoscale* **3**, 2011, 3915.
11. R. Asahi, T. Morikawa, H. Irie and T. Ohwaki, *Chem. Rev.* **114**, 2014, 9824.
12. L. Liu and X. Chen, *Chem. Rev.* **114**, 2014, 9890.
13. X. Chen, L. Liu, P. Y. Yu and S. S. Mao, *Science* **331**, 2011, 746.
14. T. Xia and X. Chen, *J. Mater. Chem. A* **1**, 2013, 2983.
15. X. Chen, L. Liu, Z. Liu, M. A. Marcus, W.-C. Wang, N. A. Oyler, M. E. Grass, B. Mao, P.-A. Glans, P. Y. Yu, J. Guo and S. S. Mao, *Sci. Rep.* **3**, 2013, 1510.
16. L. Liu, P. P. Yu, X. Chen, S. S. Mao, D. Z. Shen, *Phys. Rev. Lett.* **111**, 2013, 065505.
17. X. Chen, L. Liu and F. Huang, *Chem. Soc. Rev.* **44**, 2015, 1861.
18. C. Sun, Y. Jia, X.-H. Yang, H.-G. Yang, X. Yao, G. Q. (Max) Lu, A. Selloni and S. C. Smith, *J. Phys. Chem. C* **115**, 2011, 25590.
19. H. Lu, B. Zhao, R. Pan, J. Yao, J. Qiu, L. Luo and Y. Liu, *RSC Adv.* **4**, 2014, 1128.
20. T. Leshuk, R. Parviz, P. Everett, H. Krishnakumar, R. A. Varin and F. Gu, *ACS Appl. Mater. Interfaces* **5**, 2013, 1892.
21. T. Leshuk, S. Linley and F. Gu, *Can. J. Chem. Eng.* **91**, 2013, 799.
22. A. Naldoni, M. Allieta, S. Santangelo, M. Marelli, F. Fabbri, S. Cappelli, C. L. Bianchi, R. Psaro, and V. Dal Santo, *J. Am. Chem. Soc.* **134**, 2012, 7600.
23. M. Salari, K. Konstantinov and H. K. Liu, *J. Mater. Chem.* **21**, 2011, 5128.
24. G. Wang, H. Wang, Y. Ling, Y. Tang, X. Yang, R. C. Fitzmorris, C. Wang, J. Z. Zhang and Y. Li, *Nano Lett.* **11**, 2011, 3026.
25. W. Zhou, W. Li, J.-Q. Wang, Y. Qu, Y. Yang, Y. Xie, K. Zhang, L. Wang, H. Fu and D. Zhao, *J. Am. Chem. Soc.* **136**, 2014, 9280.
26. N. Liu, C. Schneider, D. Freitag, M. Hartmann, U. Venkatesan, J. Muller, E. Spiecker and P. Schmuki, *Nano Lett.* **14**, 2014, 3309.
27. Z. Lu, C.-T. Yip, L. Wang, H. Huang and L. Zhou, *Chem. Plus Chem.* **77**, 2012, 991.
28. A. Danon, K. Bhattacharyya, B. K. Vijayan, J. Lu, D. J. Sauter, K. A. Gray, P. C. Stair and E. Weitz, *ACS Catal.* **2**, 2012, 45.
29. Y. Zhu, D. Liu and M. Meng, *Chem. Commun.* **50**, 2014, 6049.
30. Z. Wang, C. Yang, T. Lin, H. Yin, P. Chen, D. Wan, F. Xu, F. Huang, J. Lin, X. Xie and M. Jiang, *Adv. Funct. Mater.* **23**, 2013, 5444.
31. F. Teng, M. Li, C. Gao, G. Zhang, P. Zhang, Y. Wang, L. Chen, and E. Xie, *Appl. Catal. B* **148–149**, 2014, 339.

32. Y. Yan, M. Han, A. Konkin, T. Koppe, D. Wang, T. Andreu, G. Chen, U. Vetter, J. M. Morante and P. Schaaf, *J. Mater. Chem. A* **2**, 2014, 12708.
33. H. Cui, W. Zhao, C. Yang, H. Yin, T. Lin, Y. Shan, Y. Xie, H. Gu and F. Huang, *J. Mater. Chem. A* **2**, 2014, 8612.
34. Z. Wang, C. Yang, T. Lin, H. Yin, P. Chen, D. Wan, F. Xu, F. Huang, J. Lin, X. Xie and M. Jiang, *Energy Environ. Sci.* **6**, 2013, 3007.
35. G. Zhu T. Lin, X. Lu, W. Zhao, C. Yang, Z. Wang, H. Yin, Z. Liu, F. Huang and J. Lin, *J. Mater. Chem. A* **1**, 2013, 9650.
36. C. Yang, Z. Wang, T. Lin, H. Yin, X. Lu, D. Wan, T. Xu, C. Zheng, J. Lin, F. Huang, X. Xie and M. Jiang, *J. Am. Chem. Soc.* **135**, 2013, 17831.
37. T. Lin, C. Yang, Z. Wang, H. Yin, X. Lu, F. Huang, J. Lin, X. Xie and M. Jiang, *Energy Environ. Sci.* **7**, 2014, 967.
38. H. Wang, T. Lin, G. Zhu, H. Yin, X. Lü, Y. Li and F. Huang, *Catal. Commun.* **60**, 2015, 55.
39. Q. Kang, J. Cao, Y. Zhang, L. Liu, H. Xu and J. Ye, *J. Mater. Chem. A* **1**, 2013, 5766.
40. L. R. Grabstanowicz, S. Gao, T. Li, R. M. Rickard, T. Rajh, D.-J. Liu and T. Xu, *Inorg. Chem.* **52**, 2013, 3884.
41. X. Xin, T. Xu, J. Yin, L. Wang and C. Wang, *Appl. Catal. B: Environ.* **176**, 2015, 354.
42. H. Li, Z. Chen, C. K. Tsang, Z. Li, X. Ran, C. Lee, B. Nie, L. Zheng, T. Hung, J. Lu, B. Pan and Y. Y. Li, *J. Mater. Chem. A* **2**, 2014, 229.
43. H. Zhou and Y. Zhang, *J. Phys. Chem. C* **118**, 2014, 5626.
44. L. Zheng, H. Cheng, F. Liang, S. Shu, C. K. Tsang, H. Li, S.-T. Lee and Y. Y. Li, *J. Phys. Chem. C* **116**, 2012, 5509.
45. C. Xu, Y. Song, L. Lu, C. Cheng, D. Liu, X. Fang, X. Chen, X. Zhu and D. Li, *Nanoscale Res. Lett.* **8**, 2013, 391.
46. S.-T. Myung, M. Kikuchi, C. S. Yoon, H. Yashiro, S.-J. Kim, Y.-K. Sun and B. Scrosati, *Energy Environ. Sci.* 6, 2013, 2609.
47. J. Dong, J. Han, Y. Liu, A. Nakajima, S. Matsushita, S. Wei and W. Gao, *ACS Appl. Mater. Interfaces* **6**, 2014, 1385.
48. G. Li, Z. Lian, X. Li, Y. Xu, W. Wang, D. Zhang and F. Tian, *J. Mater. Chem. A* **3**, 2015, 3748.
49. T. Xia, C. Zhang, N. A. Oyler and X. Chen, *Adv. Mater.* **25**, 2013, 6905.
50. X. Zou, J. Liu, J. Su, F. Zuo, J. Chen and P. Feng, *Chem-Eur. J.* **19**, 2013, 2866.
51. Z. Zheng, B. Huang, J. Lu, Z. Wang, X. Qin, X. Zhang, Y. Dai and M.-H. Whangbo, *Chem. Commun.* **48**, 2012, 5733.
52. J. Lu, Y. Dai, H. Jin and B. Huang, *Phys. Chem. Chem. Phys.* **13**, 2011, 18063.

53. H.-X. Deng, S.-S. Li, J. Li and S.-H. Wei, *Phys. Rev.* **85**, 2012, 195328.
54. L. Liu, Q. Liu, Y. Zheng, Z. Wang, C. Pan and W. Xiao, *J. Phys. Chem. C* **118**, 2014, 3471.
55. Y. Yan, B. Hao, D. Wang, G. Chen, E. Markweg, A. Albrecht and P. Schaaf, *J. Mater. Chem. A* **1**, 2013, 14507.

Chapter 2

Efficient Photocatalysis using Hematite Nanostructures and their Derivatives

Chun Du, James E. Thorne and Dunwei Wang

Department of Chemistry
Boston College, Merkert Chemistry Center
2609 Beacon Street, Chestnut Hill
MA 02467, USA
E-mail: dunwei.wang@bc.edu

1. Hematite for Artificial Photosynthesis

Among all of the renewable resources available to the planet, the sun has the greatest potential to fulfill the rising global energy demands. Yet, because sunlight is diffuse, diurnal, and sporadic, it remains a challenge to effectively harvest solar energy, and convert it to chemical energy. Of the existing artificial synthesis methods being studied, photoelectrochemical (PEC) solar water splitting has the potential to be an environmentally benign conversion pathway which has the advantages of being low cost, having easily configurable setup, and high conversion efficiencies. Given the abundance of water on this planet, solar water splitting stands out as a practical conversion pathway to produce hydrogen gas, which is a clean chemical fuel that can be readily stored and transported.

Water splitting was first demonstrated by Fujishima and Honda using TiO_2 under ultraviolet light.[1] Since then TiO_2,[2-5] and several other metal oxide semiconductors such as WO_3,[6,7] Fe_2O_3,[8,9] ZnO,[10-12] and Cu_2O[13-15] have been widely studied as PEC water splitting photoelectrodes. A great deal of effort has been put forth towards finding a material which has a suitable bandgap for water splitting with a high conversion efficiency. This ideal material will need to be non-toxic, inexpensive, and stable in aqueous solution, when considering its use for large scale applications. To date, no single photoelectrode material can achieve each of these critical requirements. While WO_3, α-Fe_2O_3, and Cu_2O are all promising materials, each has a bandgap that is either too small or incorrectly positioned for both spontaneous water oxidation and reduction, or both. However, research around these materials proceeds in hopes of creating a tandem cell device which utilizes two semiconductors with staggered bandgaps, one as a photoanode and the other as a photocathode.

With both Fe and O being abundant elements, with a bandgap of 2–2.2 eV, and a possible theoretical solar-to-hydrogen conversion efficiency of up to 15% with hematite, α-Fe_2O_3, or hematite, is one of the most attractive n-type semiconductor photoanode materials.[9] Currently, hematite suffers from low solar-to-hydrogen efficiencies, which is determined by light absorption, charge separation/collection, and surface catalytic activity. For hematite, the absorption coefficient, α, is relatively low, meaning that light must penetrate deeper into the film before being completely absorbed. For example, a 400 nm thick film of hematite is needed to absorb 95% of light at 550 nm.[16] Hematite also has an extremely short, excited state electron lifetime around 1 ps.[17,18] Together with its poor electrical conductivity and rapid recombination rate, this limits its minority hole carriers to a diffusion length of approximately 2–4 nm, thus hindering charge separation of bulk hematite photoanodes.[19,20] In addition, hematite shows sluggish reaction kinetics,[9] which makes an oxygen evolution reaction (OER) catalyst on surface necessary. Despite these challenges, there has been significant progress made towards the performance of hematite photoanodes through nanostructuring.

2. Nanostructured Hematite

Nanostructuring opens a new avenue of research to combat some of the aforementioned issues associated with hematite. The dimensions of nanostructures allow photogenerated carriers to more easily reach the semiconductor/electrolyte interface by decreasing the distance to the interface. Nanostructured photoelectrodes also have much larger semiconductor/electrolyte interfaces, leading to greater water oxidation rates. Also, the rough surfaces of nanostructured materials is thought to allow for a higher absorption of photons due to light scattering.[21] Each of these characteristics of nanostructured hematite promises improved charge separation and collection. Combining nanostructuring with elemental doping and surface treatments, an enormous amount of progress has been made towards making hematite a realistic photoanode for water splitting. However, high surface recombination rates remain an issue in nanostructured hematite, which will require significant future attention.[22] The following content provides an overview of the recent fabrication, characterization and performance of nanostructured hematite photoanodes, and advantages that can be gained through dopants and surface modifications.

2.1 One-dimensional nanostructures

One of the first examples of nanostructured hematite came from Hagfeldt and co-workers in 2001.[23] They provided a simple method to produce hematite nanowire arrays from a well controlled precipitation of Fe^{3+} ions in aqueous solution, which has since been extensively applied to make one-dimensional (1D) nanowire/nanorod hematite structures. This cost-effective approach of synthesizing nanostructured thin films demonstrated a method that can be envisioned for large scale applications. The details of the method include growing iron oxyhydroxide (akaganeite, β-FeOOH) on a conductive substrate in an acidic aqueous solution containing $FeCl_3$ and $NaNO_3$. A calcination was then performed in air to convert the β-FeOOH to hematite with a uniform and highly ordered nanoarchitecture. This procedure can also be applied to a variety of substrates, such as TiO_2, as seen in Figure 1.

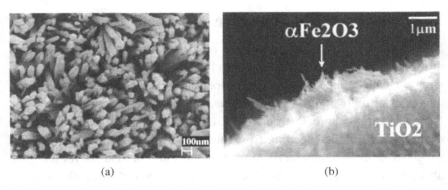

(a) (b)

Fig. 1. (a) Top view of high resolution scanning electron microscope (SEM) images of hematite nanowire grown on FTO substrates, hematite bundles are perpendicular ordered, (b) cross-section SEM image of hematite nanorods grown on TiO_2 mesoporous thin film. Figures are reproduced from Ref. 23.

Since its publication, this method has received considerable attention and several similar methods have been reported.[24,25] In 2011, Y. Li and co-authors reported a method to grow hematite nanowires with a relatively low temperature of about 350°C by means of a two-step annealing process.[26] In this method, hydrothermally grown β-FeOOH nanowires were firstly grown and then annealed in a nitrogen atmosphere at 350°C to form the magnetite phase of hematite. A subsequent annealing in air at 550°C was then used to convert the nanowires into the desired alpha hematite. This two-step heating process intentionally created oxygen vacancies to increase the electrical conductivity of hematite. As a result, an enhancement in the photocurrent density was observed using this two-step fabricated hematite compared to the one step thermal annealing of β-FeOOH nanowires at 550°C.

Another method, used by several groups to produce hematite nanowires, is a simple thermal oxidation of Fe foils in air.[27–29] Due to a volume expansion, when going from a metal to a metal oxide, arrays of Fe_2O_3 nanowires spontaneously grow from the surface of the Fe foil when heated. This method typically has high growth rates and can produce highly crystalline materials.

Aluminum oxide (AAO) templates can also be used to prepare well distributed and oriented hematite nanorod arrays.[30,31] Mao et al. were the first to use an AAO template to grow hematite nanorods. Their method firstly grew Au nanorod arrays inside AAO nanotubes. The Au nanorods provided a conductive surface, for the electrochemical deposition of a Fe layer while also acting as the current collector for the hematite photoanode. The electrodeposited Fe nanorods were then annealed in an oxygen atmosphere to convert them into a Fe_2O_3 nanorod arrays. This process is shown in Fig. 2(a), with the corresponding SEM images at each step shown Fig. 2(b).

Similar to nanowires/nanorods, nanotubes represent another well studied 1D hematite nanostructure.[31–34] In 2006, Grimes et al. reported a method to grow nanoporous hematite by potentiostatic anodization.[35] The anodization was performed using a two-electrode configuration with Fe foil as the anode and platinum foil as the counter electrode. The solution for the potentiostatic anodization contained 1% HF, 0.5% NH_4F, and 0.2% HNO_3 in glycerol. The pore size and nanotube length of hematite grown using this method can be well controlled via the anodization potential, time, and bath chemistry.

2.2 Three-dimensional (3D) hematite nanostructures

In 2010, the Grätzel group reported a porous nanostructure of hematite (Fig. 3 (a)) made through solution-based colloidal approach.[20] In this study, hematite nanoparticles were prepared by thermal decomposition of $Fe(CO)_5$ which were later dispersed in an organic solvent, and were then uniformly spread onto a conductive substrate. After this sample was annealed at 800°C, water oxidation was observed. Under 1 sun of illumination in 1 M NaOH, the photocurrent density reached 0.56 mA/cm² at 1.23 V_{RHE}. The Grätzel group later introduced a SiO_2 scaffold to help control hematite morphology and functionality.[36] After hematite was formed, the scaffold was removed and a remarkable increase in the photocurrent density from 1.57 mA/cm² to 2.34 mA/cm² was observed. This is the highest reported photocurrent from a solution processed hematite photoanode.

Other methods, such as electrodeposition, have also been widely studied to produce nanoporous hematite thin films.[37–39] A report from

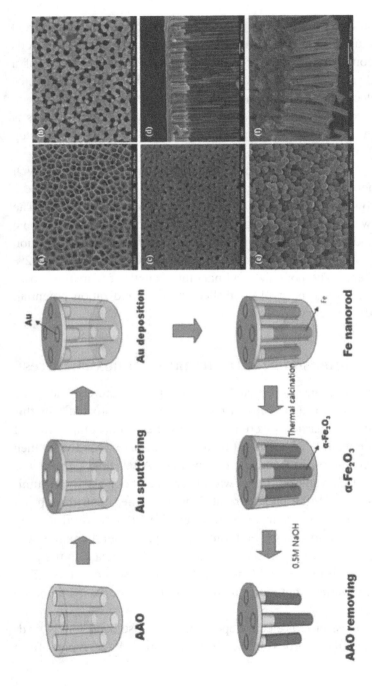

Fig. 2. Detailed steps and schematic diagrams of using AAO as templates to grow hematite nanorods. (a) Top view of the backside of pure AAO template, (b) top view of the backside of the AAO template after Au sputtering, (c) top view of the backside of the AAO template after Au nanorod growth process, (d) side view of the Au nanorods inside of the AAO template, (e) top view of Au nanorods without AAO, (f) side view of Au nanorods without AAO template. Both figures are reproduced from Ref. 30.

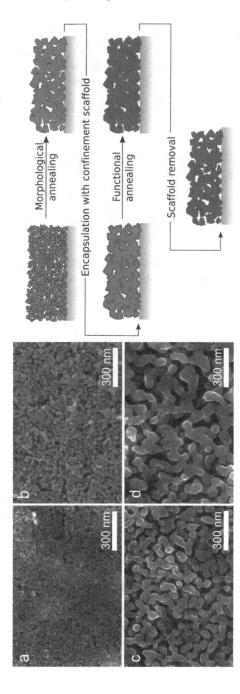

Fig. 3. (a–d). High resolution (HR) SEM images of nanoporous hematite prepared by solution colloidal method with various heat treatment parameters. (a) as deposited, (b) after 10 h at 400°C, and after 20 min at 700°C, (c) and 800°C, (d) reproduced from Ref. 20 (e) fabrication procedures of hematite nanoporous structures by using SiO$_2$ as scaffold. Reproduced from Ref. 36.

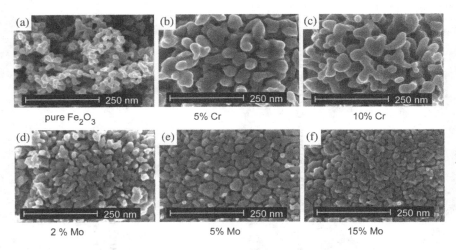

Fig. 4. Top view HR-SEM of electrodeposited samples (a) Undoped sample, (b) 15% Cr in solution, (c) 10% Cr in solution, (d) 2% Mo in solution, (e) 5% Mo in solution, and (f) 15% Mo in solution. Reproduced from Ref. 39.

McFarland *et al.* utilizes an easy way to involve extrinsic dopants, such as Cr or Mo, to improve hematite PEC behavior.[39] Using a three-electrode configuration, the electrodeposition was performed in a solution consisting of 5 mM $FeCl_3$, 5 mM KF, 0.1 M KCl, and 1 M H_2O_2. One of either $CrCl_3$ or $MoCl_5$ was used as the precursor in the electrolyte for preparing the doped samples.

The vapor phase synthesis of nanostructured hematite is another promising method that provides a quick and clean synthesis of hematite. In 2006, Grätzel *et al.* reported a high surface area nanostructured hematite with a record PEC performance grown by atmospheric pressure chemical vapor deposition (APCVD) on fluorine-doped tin oxide (FTO).[40] The HR-SEM images (Fig. 5) show that the nanostructured hematite has a thickness of about 500 nm that consist of smaller 10–20 nm features. This benchmark hematite photoelectrode, exhibited a high IPCE of 42% at 370 nm and photocurrent densities of 2.2 mA/cm² at 1.23 V_{RHE} under standard 1 sun illumination. Around the same time, the same group also reported nanostructured hematite grown by ultrasonic spray pyrolysis (USP).[41] In this method, the hematite nanostructured film turned consisted of stacked sheet that sits vertical to the substrate, as seen in HR-SEM images (Fig. 5).[42]

Efficient Photocatalysis using Hematite Nanostructures and their Derivatives 35

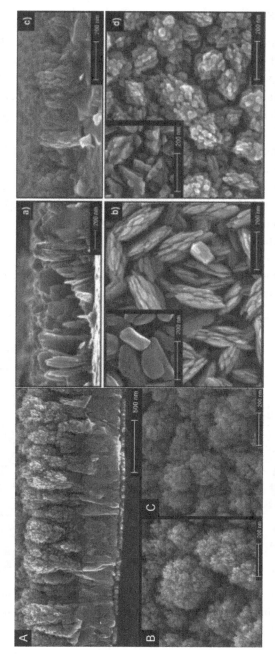

Fig. 5. Left: (a) Cross-section of 500 nm thick mesoporous Si-doped Fe_2O_3, (b) top view (45° tilted) of the Si-doped Fe_2O_3 film, (C) top view (45° tilted) of an undoped Fe_2O_3 film. Reproduced from Ref. 40. Right: Typical HR-SEM images of Si-doped hematite films on TCO obtained from USP (a,b) and APCVD (c,d); (a) and (c) are side view, (b) and (d) top view images. ((b), inset) Hematite grains for undoped USP ((d), Inset) Hematite grains for undoped APCVD electrodes. Reproduced from Ref. 42.

2.3 Hematite scaffold heteronanostructures

One alternative approach to improve the performance of hematite is to deposit a layer of hematite onto a conductive nanostructure scaffold. In this approach, the short charge diffusion of hematite can be combatted while the high surface area/high conductivity nanostructured support can shuttle collected charges more efficiently. Heteronanostructures require a high quality interface in order to facilitate charge transfer between two materials. The Wang group has used Atomic Layer Deposition (ALD) to grow uniform and conformal hematite thin films on several nanostructured scaffolds, such as 2D $TiSi_2$ nanonets, 1D Si nanowires and AZO nanotubes.[43–45] The ALD method used in the above studies represents a powerful vapor phase deposition technique which has the advantage of delivering a uniform and conformal growth on various high aspect ratio substrates.[43,46]

The 2D $TiSi_2$ nanonets, synthesized by CVD, are promising heteronanostructures that have good electronic conductivity with high surface

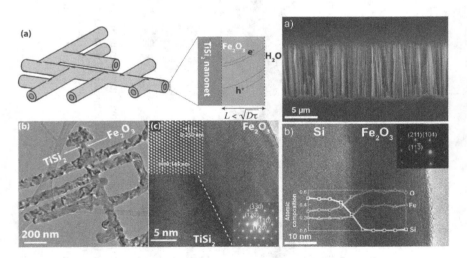

Fig. 6. SEM images display the dimension and morphology of hematite/$TiSi_2$ and hematite/SiNW. TEM images indicates that ALD technique provides a good interface between substrate and hematite. Reproduced from Refs. 43,44.

areas. Comparing the deposition of 25 nm hematite thin films on planar FTO, and 2D, TiSi$_2$ resulting in photocurrent densities of 0.41 mA/cm^2 and 1.6 mA/cm^2 at 1.23 V_{RHE}, respectively shows the improvement gained through the heteronanostructure.[43]

Si nanowires have also been used as support for hematite thin films. In this case, n-type Si nanowires act as a structural support as well as a second absorber, providing additional photovoltage. Considering the bandgap difference between Si and hematite, Si utilizes the portion of solar spectrum which is transparent to hematite. As a result, this heteronanostructure of Si/hematite exhibited an excellent turn on potential as low as 0.6 V_{RHE} which is one of the best reported values in the literature.[44]

3. Doping Hematite

An effective approach to improve electrical conductivity, thus substantially increasing photocurrents, is to incorporate dopants, such as Ti (IV),[47–51] Si (IV),[22,47,52] Sn (IV),[20,24,53–55] Mo (V)[39] Al (III)[53,56] or Pt (IV)[25,38] into hematite. Intrinsic hematite single crystals have an electrical conductivity of about $10^{-6} \Omega^{-1} cm^{-1}$.[9] This low conductivity can be explained by Fe^{3+}/Fe^{2+} valence alternation on spatially localized 3D orbitals of Fe.[9] The introduction of extrinsic dopants boosts hematite's electrical conductivity by up to several orders of magnitude. This enhancement in charge transport leads to better turn on potentials as well as improved photocurrents, as observed in the following studies involving Sn and Ti dopants.

3.1 Sn-doping

Recently, there has been a great deal of interest surrounding Sn-doped hematite nanostructures for PEC water splitting. Sn^{4+} is a well-known tetravalent dopant that can be incorporated into a hematite lattice at the Fe^{3+} sites through variety of methods.[48] Sn-doping of hematite was first reported by Shinar et al. in 1981.[57] In this study, they doped high purity hematite by mixing it with Sn followed by pressing them together and a high temperature calcination process. In 2010, Sivula et al. reported the growth of nanostructured hematite using a solution-based colloidal method while incorporating a Sn dopant.[20] These hematite electrodes were annealed at 800°C and exhibited a high photoactivity for water splitting that was attributed to the unintentional Sn^{4+} doping from FTO substrate upon the high temperature annealing process. In 2011, Li *et al.* reported a Sn-doped hematite nanocorals prepared by a simple hydrothermal method, using a Sn (IV) chloride ethanol solution as the Sn precursor.[24] This Sn-doped hematite nanostructure showed improved photocurrent densities of 0.56 mA cm^{-2} at 1.23 V_{RHE}, which was a result of the improved carrier density and structural morphology brought by Sn-doping.

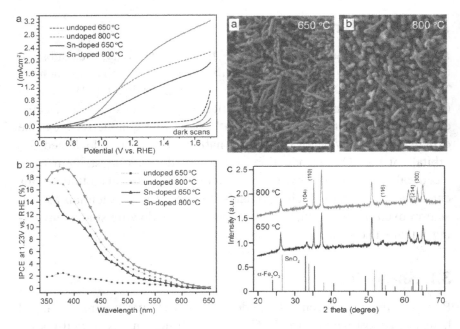

Fig. 7. (a) Comparison of J–V scans collected for hematite with and without Sn-doping sintered at 650°C and 800°C in 1 M NaOH electrolyte under 1 sun illumination, (b) comparison of IPCEs for hematite with and without Sn-doping sintered at 650°C and 800°C, collected at the incident wavelength range from 350 to 650 nm at a potential of 1.23 V vs. RHE.

Top view SEM images show the different morphologies of hematite nanowires at 650 and 800°C. Reproduced from Ref. 24.

3.2 Ti-doping

Similar to Sn-doped hematite, titanium (IV) has been successfully used as dopant for hematite and has led to improved performances. Many different methods have been used to introduce Ti dopants, such as ALD,[58] pulsed laser deposition,[59] spray pyrolysis deposition,[56] deposition-annealing process,[26] hydrothermal growths.[50,60,61] To date, there is still debate about the functioning mechanism of Ti dopant. The increase in electrical conductivity of the Ti doped hematite films may not entirely account for the large photoelectrochemical enhancements.[62] Also, the enhancement caused by Ti doping is dependent on the method of synthesis.

In one study, Hamann *et al.* reported a Ti-doped hematite thin film grown by ALD using ferrocene, water/ozone, and titanium isopropoxide precursors.[58] Taking advantage of ALD's epitaxial growth mechanism, it is possible to precisely control the morphology and thickness of hematite and concentration of Ti dopants. After comprehensive electrochemical, photoelectrochemical and impedance spectroscopy studies of undoped and Ti doped hematite electrodes, they concluded that the incorporation of an appropriate amount of Ti impurities into hematite dramatically improved hematite water oxidation performance. This enhancement was attributed to the combination of improved bulk and surface properties.

In 2012, Franking *et al.* reported a facile approach to dope hematite with Ti.[49] In this report, nanostructured hematite was synthesized by thermal oxidation of α-FeF$_3 \cdot$3H$_2$O NWs in ambient environment and titanium *n*-butoxide [Ti(OBu)$_4$] in ethanol was used as Ti precursor in this approach. As expected, they found that the Ti-treatment increased the donor concentration of hematite by 10 times, which can greatly promote majority carrier transport. Using electrochemical impedance spectroscopy (EIS) measurements, under illumination, and Mott–Schottky analysis, they concluded that Ti dopants can passivate surface trap states of hematite.

3.3 Oxygen vacancies

Creating oxygen vacancies in hematite has been reported by several groups as an alternative approach to improving the conductivity of

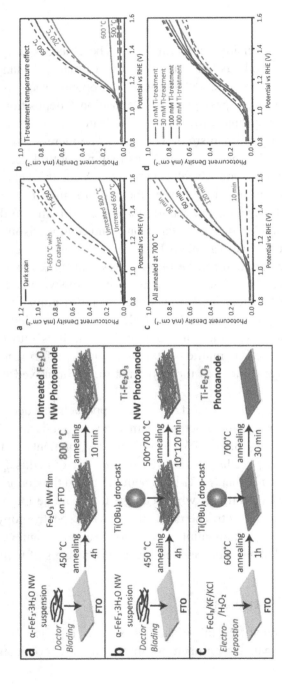

Fig. 8. Preparation schematics for: (a) an untreated hematite NW photoanode, (b) a Ti-treated hematite NW photoanode, (c) a Ti-treated electrodeposited hematite film photoanode. Photocurrent densities of hematite NW films with and without Ti doping measured under 1 sun condition. Different annealing temperatures lead to slightly different photocurrent densities. Reproduced from Ref. 49.

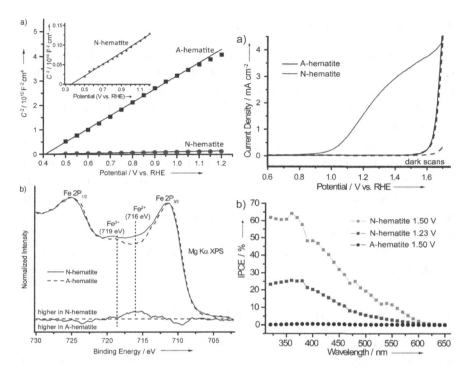

Fig. 9. Top: (a) Mott–Schottky plots measured for A-hematite and N-hematite nanowire films, (b) overlay of Fe 2p XPS spectra of A-hematite and N-hematite films. The vertical dashed lines highlight the satellite peaks for Fe^{3+} and Fe^{2+}. Bottom: (a) Linear sweep voltammograms collected for A-hematite and N-hematite in a NaOH electrolyte solution, (b) Corresponding IPCE spectra for A-hematite and N-hematite, collected at potentials of 1.23 and 1.50 V vs. RHE. Reproduced from Ref. 64.

hematite.[63–65] Hematite with high oxygen vacancies can be obtained through post annealing β-FeOOH in oxygen deficient conditions.[64] Specifically, the increase in Fe^{2+} lattice sites significantly improves the hematite electronic conductivity through polaron hopping mechanism.[66] The increase of Fe^{2+} sites in hematite has been confirmed by XPS analysis, while Mott–Schottky measurements showed an increase in carrier concentration, and the photocurrent, density of the oxygen deficient hematite, being as high as 3.37 mA/cm² at 1.50 V_{RHE}.

4. Surface Treatments of Hematite

The semiconductor/electrolyte interface is a determining feature of the reaction kinetics as well as thermodynamics in a PEC cell. For this reason, a large number of studies have focused on controlling this interface through surface modifications. One such surface modification, a passivation layer, has been widely used to suppress surface states as well as catalyze surface chemical reactions. Passivation layers can be deposited using a variety of methods, such as ALD, electrochemical deposition, dip coating and sputtering.[67–70] The application of a passivation layer has been seen to cathodically shift the turn on potential of hematite, as shown in Fig. 10 (a) Schematic depictions of the removal of surface states are shown in Figs. 10 (b–d). In the following section, some recent work on surface treatments of nanostructured hematite will be discussed.

4.1 Surface modification with OER catalyst

It is well recognized that sluggish water oxidation reaction kinetics limit the PEC performance of hematite.[71,72] Because of this, even though photogenerated charges can be effectively separated and travel to interface, they will accumulate at surface leading to a late turn on potentials. The Hamann group studied hematite reaction kinetics in aqueous electrolyte with and without a hole scavenger.[73] They concluded that the hole scavenger alleviated the overpotential needed to initiate water oxidation, indicating that the slow surface kinetics are responsible for a large portion of the needed overpotential, as seen in Fig. 11. EIS and intensity modulated photocurrent spectroscopy (IMPS) studies performed by Peter *et al.* indicate a competition of the fate of holes between charge transfer from surface states to electrolyte and recombination with conduction band electrons at surface states.[74,75] It has also been found that turn-on potentials can be affected by Fermi level pinning which is related to the magnitude of the Helmholtz layer.[74] These studies have determined that water oxidation is a rate determining step, and thus a great deal of effort has been directed into the study of surface passivation catalyst.

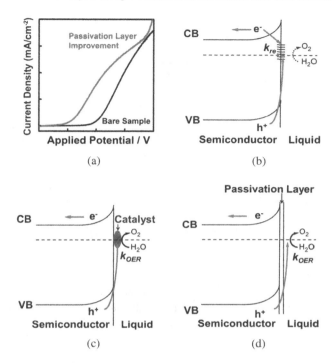

Fig. 10. (a) comparison of J–V behavior of hematite with and without surface passivation layer, (b) surface states are sitting between conduction band and valence band which can trap photogenerated charges. These states greatly hinder the reaction kinetics at semiconductor/electrolyte interface which can be restrained by surface passivation layer, (c) and (d), reproduced from Ref. 70.

Among noble metal oxides, iridium oxide is one of the best electrocatalysts for lowering the overpotential for the water oxidation reaction.[76,77] The Grätzel group achieved over 3 mA/cm² of photocurrent density at 1.23 V_{RHE} under 1 sun of illumination when applying IrO_2 nanoparticles to a nanostructured hematite surface.[78] This work represents a benchmark for nanostructured hematite with an OER catalyst for PEC water splitting system.

Compared to Ru and Ir catalysts for water oxidation, cobalt containing catalysts represent a group of earth-abundant candidates, which have attracted significant research attention. Coupling hematite with a Co catalyst has been separately reported by Choi and Grätzel groups.[79,80]

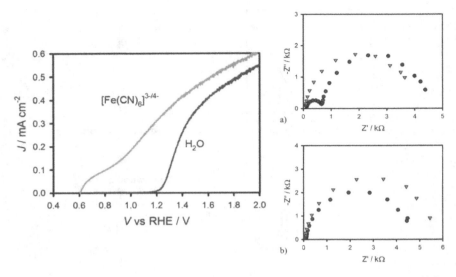

Fig. 11. (a) $J-V$ curve of a hematite electrode in contact with and without $[Fe(CN)_6]^{3-/4-}$, (b) Nyquist plots for IS data measured under 1 sun illumination for H_2O and $[Fe(CN)_6]^{3-/4-}$ electrolytes at (a) 1.3 V_{RHE} and (b) 1.6 V_{RHE}. Reproduced from Ref. 73.

Riha et al. reported a significant enhancement in PEC water splitting behavior after the ALD of a submonolayer $Co(OH)_2/Co_3O_4$ onto a nanostructured inverse opal scaffold of hematite.[81] In 2009, the Gamelin group reported the use of a Co–Pi catalyst on a hematite photoanode for the first time.[82] The Co–Pi catalyst was electrodeposited on hematite in a potassium phosphate buffer solution containing $Co(NO_3)_2$. After the deposition of Co–Pi, the onset potential shifted cathodically almost to 300 mV. Later, they optimized the deposition method by introducing a photo-assisted electrodeposition, which provided a more uniform layer of Co–Pi on to hematite.[83] This resulted in a stable photocurrent density of 2.8 mA/cm² at 1.23 V_{RHE}.

The role of the Co–Pi as an electrocatalyst has been a topic of debate.[67,84,85] While Durrant group has implied a non-catalytic role for Co-Pi, the Hamann group has suggested that Co-Pi improves PEC performance by reducing surface electron and hole recombination. The lack of a definitive conclusion about the role of Co–Pi, shows that a further

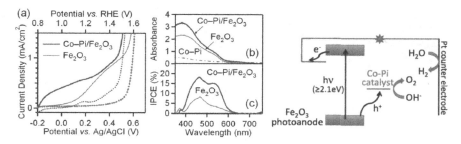

Fig. 12. (a) Dark (dashed) and photocurrent densities for Fe_2O_3 and Co–Pi/Fe_2O_3 photoanodes, collected using simulated AM1.5 illumination, (b) electronic absorption and (c) IPCE spectra for e_2O_3 and Co–Pi/ Fe_2O_3 at 1.23 and 1 V_{RHE}, respectively. Scheme of working mechanism of Co–Pi decorated hematite. Reproduced from Ref. 82.

exploration to gain a deeper understanding of OER catalysts in PEC water splitting systems is needed.

Another example of an OER catalyst on a hematite photoelectrode was reported by Du *et al.* with the use of amorphous $NiFeO_x$ as a surface modification layer.[86] $NiFeO_x$ was casted onto the hematite electrode surface via photochemical metal–organic deposition, as previously reported by Smith *et al.*[87] Compared with the bare sample ($V_{on} = 1.0$ V_{RHE}), $NiFeO_x$ modified hematite showed an early onset potential of around 0.62 V_{RHE}, which is one of the best onset potentials reported in the literature. As seen in Fig. 13, Du *et al.* were able to assign this low onset potential to a greater photovoltage produced at hematite/electrolyte interface, rather than solely increased reaction kinetics. This study shows the importance of considering both the reaction kinetics and the surface thermodynamics when applying OER catalysts to a hematite surface.

4.2 Non-catalytic surface passivation layers

Surface recombination limits the efficient utilization of photogenerated charges in the case of hematite.[68,88] Detailed studies have been performed to prove the existence of surface states and their role as the main recombination centers on hematite surface.[89,90] The application of a surface

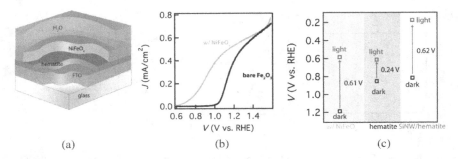

Fig. 13. (a) Schematic perspective view of the modification, the cut-through revealing various layers, (b) a substantial cathodic shift, up to 380 mV, is observed in the photocurrent density–voltage plot, (c) significant difference is obvious between the measured phovoltages by bare hematite and that with Ni decorations. As a control, the voltages measured on the SiNW/hematite system are shown on the right. Reproduced from Ref. 87.

passivation layer has appeared to be a valid way to overcome these recombination centers, where Formal *et al.* reported a decrease in overpotential when coating nanostructured hematite with a ultra-thin layer of Al_2O_3 grown by ALD.[80] This less than 2 nm Al_2O_3 overlayer shifted the onset potential cathodically by as much as 100 mV and increased the photocurrent by a factor of 3.5. Transient photocurrent spectroscopy, EIS and photoluminescence spectra were used to confirm the passivation of surface trapping states with the help of Al_2O_3. Similar improvements have been observed when using Ga_2O_3 overlayer on hematite, as seen in a study from the same group.[91] The Durrant group studied the CoO_x and Ga_2O_3 surface modifications on hematite using transient absorption investigations.[92] They indicated that the shift in the turn on potential was due to long-lived photoholes, which was the result of the passivation of electron/hole recombination. Similarly, ZnO overlayers have been used as a surface treatment layer which shifted the flat band potential of hematite and reduced surface defects.[93]

Xi *et al.* reported a novel surface treatment using a $Fe_xSn_{1-x}O_4$ layer on the surface of hematite nanorods.[94] This surface layer was formed by annealing at 750°C after coating FeOOH nanorods with a Sn (IV) aqueous solution. The improvement in photocurrent and photon conversion efficiency was attributed to the suppression of recombination at hematite/electrolyte interface. Zhang *et al.* reported a facile surface passivation method by scanning hematite in aqueous solution containing NaCl.[95] The enhancement in IPCE as well as photocurrent density was credited to the suppression of surface recombination, which was supported by current transients and photoluminescence (PL) spectra studies.

4.3 Other surface modifications

Lin *et al.* have successfully grown a thin layer of p-type Fe_2O_3 on the surface of n-type hematite by involving Mg-doping in atomic layer deposition (ALD) growth.[96] This n–p hematite junction showed a 200 mV cathodic shift in onset potential compared to that of bare n-type hematite under 1 sun illumination. This enhancement in onset potential was thought to come from the built-in potential brought by the homojunction within hematite.

Li *et al.* reported a H_2-treated hematite photoelectrode through the simple pyrolysis of $NaBH_4$.[97] After H_2 treatment, their hematite sample showed a photocurrent density of 2.28 mA/cm^2 at 1.23 V_{RHE}, which was 2.5 times higher than that of pristine hematite, and enhanced onset potentials. The authors attributed this improvement to surface oxygen deficiencies caused by the H_2 treatment. This was supported by the surface-sensitive O K-edge and bulk sensitive Fe K-edge X-ray absorption spectroscopy (XAS) data, where the surface-sensitive O K-edge XAS results indicated that the H_2 treatment generated Fe^{2+} on the surface of the hematite with oxygen vacancies. Bulk sensitive Fe K-edge XAS data showed almost identical curves for H_2-treated hematite and pristine hematite, as shown in figure 14.

To summarize, a great deal of attention has been given to hematite and in return a great deal of progress has been made. Taking advantage of nanotechniques for hematite growth and surface passivation promise to lead to even greater photocurrents, can be achieved resulting in higher solar-to-fuel conversion efficiency.

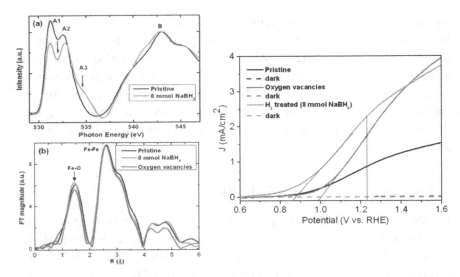

Fig. 14. (a) O K-edge XAS spectra of pristine and H_2-treated hematite nanostructures, (b) Fourier transform of the extended X-ray absorption fine structure (EXAFS) data at the Fe K-edge of pristine hematite, H_2-treated hematite nanostructures and hematite nanostructures with typical oxygen vacancies treated in an oxygen deficient atmosphere, (C) J–V scans for pristine hematite nanostructures; hematite nanostructures with typical oxygen vacancies treated in an oxygen deficient atmosphere during the sintering process and H_2-treated hematite nanostructures. Reproduced from Ref. 97.

References

1. A. Fujishima. Electrochemical photolysis of water at a semiconductor electrode. *Nature* **238**, 1972, 37–38.
2. J. H. Park, S. Kim and A. J. Bard. Novel carbon-doped TiO2 nanotube arrays with high aspect ratios for efficient solar water splitting. *Nano Lett.* **6**, 2005, 24–28.
3. C. Burda *et al*. Enhanced nitrogen doping in tio2 nanoparticles. *Nano Lett.* **3**, 2003, 1049–1051.
4. Y. J. Hwang, A. Boukai and P. Yang. High density n-Si/n-TiO2 core/shell nanowire arrays with enhanced photoactivity. *Nano Lett.* **9**, 2008, 410–415.
5. M. Ni, M. K. Leung, D. Y. Leung and K. Sumathy. A review and recent developments in photocatalytic water-splitting using TiO2 for hydrogen production. *Renew. Sust. Energ. Rev.* **11**, 2007, 401–425.

6. C. Santato, M. Ulmann and J. Augustynski. Photoelectrochemical properties of nanostructured tungsten trioxide films. *J. Phys. Chem. B* **105**, 2001, 936–940.
7. I. Bedja, S. Hotchandani and P. V. Kamat. Photoelectrochemistry of quantized tungsten trioxide colloids: electron storage, electrochromic, and photoelectrochromic effects. *J. Phys. Chem.* **97**, 1993, 11064–11070.
8. S. U. M. Khan and J. Akikusa. Photoelectrochemical splitting of water at nanocrystalline n-Fe2O3 thin-film electrodes. *J. Phys. Chem. B* **103**, 1999, 7184–7189.
9. K. Sivula, F. Le Formal and M. Grätzel. Solar Water splitting: progress using hematite (αfe_2o_3) photoelectrodes. *Chem. Sus. Chem.* **4**, 2011, 432–449.
10. K. Keis *et al.* Nanostructured ZnO electrodes for dye-sensitized solar cell applications. *J. Photoch. Photobio. A* **148**, 2002, 57–64.
11. A. Wolcott, W. A. Smith, T. R. Kuykendall, Y. Zhao and J. Z. Zhang. Photoelectrochemical study of nanostructured ZnO thin films for hydrogen generation from water splitting. *Advanced Functional Materials* **19**, 2009, 1849–1856.
12. K.-S. Ahn *et al.* ZnO nanocoral structures for photoelectrochemical cells. *Appl. Phys. Lett.* **93**, 2008, 163117.
13. A. Paracchino, V. Laporte, K. Sivula, M. Grätzel and E. Thimsen. Highly active oxide photocathode for photoelectrochemical water reduction. *Nature Mater.* **10**, 2011, 456–461.
14. A. Paracchino *et al.* Ultrathin films on copper (I) oxide water splitting photocathodes: a study on performance and stability. *Energy Environ. Sci.* **5**, 2012, 8673–8681.
15. T. Mahalingam *et al.* Photoelectrochemical solar cell studies on electroplated cuprous oxide thin films. *J. Mater. Sci-Mater. El.* **17**, 2006, 519–523.
16. T. W. Hamann. Splitting water with rust: hematite photoelectrochemistry. *Dalton Trans.* **41**, 2012, 7830–7834.
17. N. J. Cherepy, D. B. Liston, J. A. Lovejoy, H. Deng and J. Z. Zhang. Ultrafast studies of photoexcited electron dynamics in γ-and α-Fe_2O_3 semiconductor nanoparticles. *J. Phys. Chem. B* **102**, 1998, 770–776.
18. U. Bjoerksten, J. Moser and M. Graetzel. Photoelectrochemical studies on nanocrystalline hematite films. *Chem. Mater.* **6**, 1994, 858–863.
19. J. H. Kennedy and K. W. Frese. Photooxidation of Water at α-Fe2O3 Electrodes. *J. Electrochem. Soc.* **125**, 1978, 709–714.
20. K. Sivula *et al.* Photoelectrochemical water splitting with mesoporous hematite prepared by a solution-based colloidal approach. *J. Am. Chem. Soc.* **132**, 2010, 7436–7444.

21. F. E. Osterloh. Inorganic nanostructures for photoelectrochemical and photocatalytic water splitting. *Chem. Soc. Rev.* **42**, 2013, 2294–2320.
22. I. Cesar, K. Sivula, A. Kay, R. Zboril and M. Grätzel. Influence of feature size, film thickness, and silicon doping on the performance of nanostructured hematite photoanodes for solar water splitting. *J. Phys. Chem. C* **113**, 2008, 772–782.
23. L. Vayssieres, N. Beermann, S.-E. Lindquist and A. Hagfeldt. Controlled aqueous chemical growth of oriented three-dimensional crystalline nanorod arrays: application to iron (III) oxides. *Chem. Mater.* **13**, 2001, 233–235.
24. Y. Ling, G. Wang, Wheeler, D. A., J. Z. Zhang and Y. Li. Sn-doped hematite nanostructures for photoelectrochemical water splitting. *Nano Lett.* **11**, 2011, 2119–2125.
25. J. Y. Kim *et al.* Single-crystalline, wormlike hematite photoanodes for efficient solar water splitting. *Sci. Rep.* **3**, 2013.
26. Wang, G *et al.* Facile Synthesis of Highly photoactive α-Fe_2O_3-based films for water oxidation. *Nano Lett.* **11**, 2011, 3503–3509.
27. Y. Fu, J. Chen and H. Zhang. Synthesis of Fe_2O_3 nanowires by oxidation of iron. *Chem. Phys. Lett.* **350**, 2001, 491–494.
28. R. Wang, Y. Chen, Y. Fu, H. Zhang and C. Kisielowski. Bicrystalline Hematite Nanowires. *J.Phys. Chem. B* **109**, 2005, 12245–12249.
29. X. Wen, S. Wang, Y. Ding, Z. L. Wang and S. Yang. Controlled Growth of Large-Area, Uniform, Vertically Aligned Arrays of α-Fe_2O_3 Nanobelts and Nanowires. *J. Phys. Chem. B* **109**, 215–220 (2004).
30. A. Mao, G. Y. Han and J. H. Park. Synthesis and photoelectrochemical cell properties of vertically grown α-Fe_2O_3 nanorod arrays on a gold nanorod substrate. *J. Mater. Chem.* **20**, 2010, 2247–2250.
31. A. Mao *et al.* Controlled synthesis of vertically aligned hematite on conducting substrate for photoelectrochemical cells: Nanorods versus nanotubes. *ACS Appl. Mater. Interfaces* **3**, 2011, 1852–1858.
32. R. Rangaraju, A. Panday, K. Raja and M. Misra. Nanostructured anodic iron oxide film as photoanode for water oxidation. *J. Phys. D* **42**, 2009, 135303.
33. T. J. LaTempa, X. Feng, M. Paulose and C. A. Grimes. Temperature-dependent growth of self-assembled hematite (α-Fe_2O_3) nanotube arrays: rapid electrochemical synthesis and photoelectrochemical properties. *J. Phys. Chem. C* **113**, 2009, 16293–16298.
34. S. K. Mohapatra, S. E. John, S. Banerjee and M. Misra. Water Photooxidation by Smooth and Ultrathin α-Fe_2O_3 Nanotube Arrays. *Chem. Mater.* **21**, 2009, 3048–3055.

35. H. E. Prakasam, O. K. Varghese, M. Paulose, G. K. Mor and C. A. Grimes. Synthesis and photoelectrochemical properties of nanoporous iron (III) oxide by potentiostatic anodization. *Nanotechnology* **17**, 2006, 4285.
36. J. Brillet, M. Gratzel and K. Sivula. Decoupling feature size and functionality in solution-processed, porous hematite electrodes for solar water splitting. *Nano Lett.* **10**, 2010, 4155–4160.
37. R. L. Spray and K.-S. Choi. Photoactivity of transparent nanocrystalline Fe_2O_3 electrodes prepared via anodic electrodeposition. *Chemistry of Materials* **21**, 2009, 3701–3709.
38. Y.-S. Hu *et al*. Pt-Doped α-Fe_2O_3 thin films active for photoelectrochemical water splitting. *Chem. Mater.* **20**, 2008, 3803–3805.
39. A. H. Kleiman, Y.-S. Shwarstein, A. J. Forman, G. D. Stucky and E. W. McFarland. Electrodeposition of α-Fe_2O_3 Doped with Mo or Cr as Photoanodes for Photocatalytic Water Splitting. *J. Phys. Chem. C* **112**, 2008, 15900–15907.
40. A. Kay, I. Cesar and M. Grätzel. New Benchmark for Water Photooxidation by Nanostructured α-Fe_2O_3 Films. *J. Am. Chem. Soc.* **128**, 2006, 15714–15721.
41. A. Duret and M. Grätzel. Visible light-induced water oxidation on mesoscopic α-Fe_2O_3 films made by ultrasonic spray pyrolysis. *J. Phys. Chem. B* **109**, 2005, 17184–17191.
42. I. Cesar, A. Kay, J. A. Gonzalez, Martinez and M. Grätzel. Translucent thin film Fe_2O_3 photoanodes for efficient water splitting by sunlight: nanostructure-directing effect of Si-doping. *J. Am. Chem. Soc.* **128**, 2006, 4582–4583.
43. Y. Lin, S. Zhou, S. W. Sheehan and D. Wang. Nanonet-based hematite heteronanostructures for efficient solar water splitting. *J. Am. Chem. Soc.* **133**, 2011, 2398–2401.
44. M. T. Mayer, C. Du and D. Wang. Hematite/Si nanowire dual-absorber system for photoelectrochemical water splitting at low applied potentials. *J. Am. Chem. Soc.* **134**, 2012, 12406–12409.
45. Y. Lin, G. Yuan, S. Sheehan, S. Zhou and D. Wang. Hematite-based solar water splitting: challenges and opportunities. *Energy Environ. Sci.* **4**, 2011, 4862–4869.
46. B. M. Klahr, A. B. Martinson and T. W. Hamann. Photoelectrochemical investigation of ultrathin film iron oxide solar cells prepared by atomic layer deposition. *Langmuir* **27**, 2010, 461–468.
47. J. A. Glasscock, P. R. Barnes, I. C. Plumb and N. Savvides. Enhancement of photoelectrochemical hydrogen production from hematite thin films by the introduction of Ti and Si. *J. Phys. Chem. C* **111**, 2007, 16477–16488.

48. N. T. Hahn and C. B. Mullins. Photoelectrochemical performance of nanostructured Ti-and Sn-doped α-Fe$_2$O$_3$ photoanodes. *Chem. Mater.* **22**, 2010, 6474–6482.
49. R. Franking *et al*. Facile post-growth doping of nanostructured hematite photoanodes for enhanced photoelectrochemical water oxidation. *Energy Environ. Sci.* **6**, 2013, 500–512.
50. S. Shen *et al*. Physical and photoelectrochemical characterization of Ti-doped hematite photoanodes prepared by solution growth. *J. Mater. Chem. A* **1**, 2013, 14498–14506.
51. J. Liu, C. Liang, H. Zhang, Tian, Z and S. Zhang. General strategy for doping impurities (Ge, Si, Mn, Sn, Ti) in hematite nanocrystals. *J. Phys. Chem. C* **116**, 2012, 4986–4992.
52. Y. Liang, C. S. Enache and R. van de Krol. Photoelectrochemical characterization of sprayed α-Fe$_2$O$_3$ thin films: Influence of Si Doping and SnO$_2$ Interfacial Layer. *Int. J. Photoenergy* **2008**, 2008.
53. J. S. Jang, J. Lee, H. Ye, F.-R. F. Fan and A. J. Bard. Rapid screening of effective dopants for Fe$_2$O$_3$ photocatalysts with scanning electrochemical microscopy and investigation of their photoelectrochemical properties. *J. Phys. Chem. C* **113**, 2009, 6719–6724.
54. J. Frydrych *et al*. Facile fabrication of tin-doped hematite photoelectrodes–effect of doping on magnetic properties and performance for light-induced water splitting. *J. Mater. Chem.* **22**, 2012, 23232–23239.
55. J. S. Jang, K. Y. Yoon, X. Xiao, F.-R. F. Fan and A. J. Bard. Development of a potential Fe2O3-based photocatalyst thin film for water oxidation by scanning electrochemical Microscopy: Effects of Ag– Fe$_2$O$_3$ nanocomposite and Sn Doping. *Chem. Mater.* **21**, 2009, 4803–4810.
56. C. Jorand Sartoretti *et al*. Photoelectrochemical oxidation of water at transparent ferric oxide film electrodes. *J. Phys. Chem. B* **109**, 2005, 13685–13692.
57. J. H. Kennedy, M. Anderman and R. Shinar. Photoactivity of polycrystalline α-Fe$_2$O$_3$ electrodes doped with Group IVA elements. *J. Electrochem. Soc.* **128**, 1981, 2371–2373.
58. O. Zandi, B. M. Klahr and T. W. Hamann. Highly photoactive Ti-doped [small alpha]-Fe$_2$O$_3$ thin film electrodes: resurrection of the dead layer. *Energy Environ. Sci.* **6**, 2013, 634–642.
59. C. X. Kronawitter, S. S. Mao and B. R. Antoun. Doped, porous iron oxide films and their optical functions and anodic photocurrents for solar water splitting. *Appl. Phys. Lett.* **98**, 2011, 092108.

60. J. Deng *et al.* Ti-doped hematite nanostructures for solar water splitting with high efficiency. *J. Appl. Phys.* **112**, 2012, 084312.
61. Miao, C. *et al.* Micro-Nano-Structured Fe_2O_3: Ti/ZnFe2O4 Heterojunction Films for Water Oxidation. *ACS Appl. Mater. Interfaces* **4**, 2012, 4428–4433.
62. C. Kronawitter *et al.* Titanium incorporation into hematite photoelectrodes: Theoretical considerations and experimental observations. *Energy Environ. Sci.* **7**, 2014, 3100–3121.
63. G. Wang, Ling, Y and Li, Y. Oxygen-deficient metal oxide nanostructures for photoelectrochemical water oxidation and other applications. *Nanoscale* **4**, 2012, 6682–6691.
64. Y. Ling *et al.* The influence of oxygen content on the thermal activation of hematite nanowires. *Angewandte Chemie* **124**, 2012, 4150–4155.
65. A. Pu *et al.* Coupling Ti-doping and oxygen vacancies in hematite nanostructures for solar water oxidation with high efficiency. *J. Mater. Chem. A* **2**, 2014, 2491–2497.
66. N. Iordanova, Dupuis, M and K. M. Rosso. Charge transport in metal oxides: a theoretical study of hematite α-Fe_2O_3. *J. Chem. Phys.* **122**, 2005, 144305.
67. B. Klahr, S. Gimenez, F. Fabregat-Santiago, J. Bisquert and T. W. Hamann. Photoelectrochemical and Impedance Spectroscopic Investigation of Water Oxidation with "Co–Pi"-Coated Hematite Electrodes. *J. Am. Chem. Soc.* **134**, 2012, 16693–16700.
68. F. Le Formal *et al.* Back Electron–Hole Recombination in Hematite Photoanodes for Water Splitting. *J. Am. Chem. Soc.* **136**, 2014, 2564–2574.
69. K. Sivula. Metal oxide photoelectrodes for solar fuel production, surface traps, and catalysis. *J. Phys. Chem. Lett* **4**, 2013, 1624–1633.
70. R. Liu, Z. Zheng, J. Spurgeon, B. S. Brunschwig and X. Yang. Enhanced Photoelectrochemical Water-Splitting Performance of Semiconductors by Surface Passivation Layers. *Energy Environ. Sci.* 2014.
71. K. L. Hardee and A. J. Bard. Semiconductor electrodes V. The application of chemically vapor deposited iron oxide films to photosensitized electrolysis. *J. Electrochem. Soc.* **123**, 1976, 1024–1026.
72. H. Dotan, K. Sivula, M. Gratzel, Rothschild, A and S. C. Warren. Probing the photoelectrochemical properties of hematite ([small alpha]-Fe_2O_3) electrodes using hydrogen peroxide as a hole scavenger. *Energy Environ. Sci.* **4**, 2011, 958–964.
73. B. Klahr, S. Gimenez, F. Fabregat-Santiago, J. Bisquert and T. W. Hamann. Electrochemical and photoelectrochemical investigation of water oxidation with hematite electrodes. *Energy Environ. Sci.* **5**, 2012, 7626–7636.

74. K. G. Upul Wijayantha, S. Saremi-Yarahmadi and L. M. Peter. Kinetics of oxygen evolution at [small alpha]-Fe2O3 photoanodes: a study by photoelectrochemical impedance spectroscopy. *Phys. Chem. Chem. Phys.* **13**, 2011, 5264–5270.
75. L. M. Peter. Energetics and kinetics of light-driven oxygen evolution at semiconductor electrodes: the example of hematite. *J. Solid State Electr.* **17**, 2013, 315–326.
76. J. Kiwi and M. Grätzel. Colloidal Redox Catalysts for Evolution of Oxygen and for Light-Induced Evolution of Hydrogen from Water. *Angew. Chem. Int. Edit.* **18**, 1979, 624–626.
77. T. Reier, M. Oezaslan and P. Strasser. Electrocatalytic oxygen evolution reaction (OER) on Ru, Ir, and Pt catalysts: a comparative study of nanoparticles and bulk materials. *ACS Catalysis* **2**, 2012, 1765–1772.
78. S. D. Tilley, M. Cornuz, K. Sivula and M. Grätzel. Light-Induced water splitting with hematite: Improved nanostructure and iridium oxide catalysis. *Angew. Chem.* **122**, 2010, 6549–6552.
79. K. J. McDonald and K.-S. Choi. Synthesis and photoelectrochemical properties of $Fe_2O_3/ZnFe_2O_4$ composite photoanodes for use in solar water oxidation. *Chem. Mater.* **23**, 2011, 4863–4869.
80. F. Le Formal *et al.* Passivating surface states on water splitting hematite photoanodes with alumina overlayers. *Chemical Science* **2**, 2011, 737–743.
81. S. C. Riha *et al.* Atomic layer deposition of a submonolayer catalyst for the enhanced photoelectrochemical performance of water oxidation with hematite. *ACS Nano* **7**, 2013, 2396–2405.
82. D. K. Zhong, J. Sun, H. Inumaru and D. R. Gamelin. Solar Water Oxidation by Composite Catalyst/α-Fe_2O_3 Photoanodes. *J. Am. Chem. Soc.* **131**, 2009, 6086–6087.
83. D. K. Zhong, M. Cornuz, K. Sivula, M. Grätzel and D. R. Gamelin. Photo-assisted electrodeposition of cobalt–phosphate (Co–Pi) catalyst on hematite photoanodes for solar water oxidation. *Energy Environ. Sci.* **4**, 2011, 1759–1764.
84. D. R. Gamelin. Water splitting: Catalyst or spectator? *Nature Chem.* **4**, 2012, 965–967.
85. M. Barroso *et al.* The role of cobalt phosphate in enhancing the photocatalytic activity of α-Fe2O3 toward water oxidation. *J. Am. Chem. Soc.* **133**, 2011, 14868–14871.
86. C. Du *et al.* Hematite-Based Water Splitting with Low Turn-On Voltages. *Angew. Chem. Int. Ed.* **52**, 2013, 12692–12695.

87. R. D. Smith *et al*. Photochemical route for accessing amorphous metal oxide materials for water oxidation catalysis. *Science* **340**, 2013, 60–63.
88. M. Barroso, S. R. Pendlebury, A. J. Cowan and J. R. Durrant. Charge carrier trapping, recombination and transfer in hematite ([small alpha]-Fe$_2$O$_3$) water splitting photoanodes. *Chemical Science* **4**, 2013, 2724–2734.
89. B. Klahr, S. Gimenez, F. Fabregat-Santiago, T. Hamann and J. Bisquert. Water oxidation at hematite photoelectrodes: the role of surface states. *J. Am. Chem. Soc.* **134**, 2012, 4294–4302.
90. C. Du *et al*. Observation and Alteration of Surface States of Hematite Photoelectrodes. *J. Phys. Chem. C* 2014.
91. T. Hisatomi *et al*. Cathodic shift in onset potential of solar oxygen evolution on hematite by 13-group oxide overlayers. *Energy Environ. Sci.* **4**, 2011, 2512–2515.
92. M. Barroso *et al*. Dynamics of photogenerated holes in surface modified α-Fe$_2$O$_3$ photoanodes for solar water splitting. *Proc. Natl. Acad. Sci.* **109**, 2012, 15640–15645.
93. P. SaurabháBassi, S. YangáChiam, W. FattáMak, J. S. Loo and L. HelenaáWong. Surface treatment of hematite photoanodes with zinc acetate for water oxidation. *Nanoscale* **4**, 2012, 4430–4433.
94. L. Xi *et al*. A novel strategy for surface treatment on hematite photoanode for efficient water oxidation. *Chemical Science* **4**, 2013, 164–169.
95. M. Zhang *et al*. A facile strategy to passivate surface states on the undoped hematite photoanode for water splitting. *Electrochem. Commun.* **23**, 2012, 41–43.
96. Y. Lin *et al*. Growth of p-type hematite by atomic layer deposition and its utilization for improved solar water splitting. *J. Am. Chem. Soc.* **134**, 2012, 5508–5511.
97. M. Li *et al*. Hydrogen-treated hematite nanostructures with low onset potential for highly efficient solar water oxidation. *J. Mater. Chem. A* **2**, 2014, 6727–6733.

Chapter 3

One-Dimensional Silicon Nanowire Composites for Photocatalysis

Yuanyuan Ma[*], Jiayuan Li[*] and Yongquan Qu,[*,†,‡]

[*]*Center for Applied Chemical Research,
Frontier Institute of Science and Technology and
State Key Laboratory for Mechanical Behavior of Materials
Xi'an Jiaotong University, Xi'an 710049, PRC*
[†]*MOE Key Laboratory for Non-equilibrium Synthesis and
Modulation of Condensed Matter Xi'an
Jiaotong University, Xi'an 710049, PRC*
[‡]*yongquan@mail.xjtu.edu.cn*

1. Introduction

Silicon, as the second most abundant element on Earth, is the most dominant semiconductor for various electronic applications in the modern industries. Among various applications, silicon represents a high efficient material for solar energy conversion, in which silicon-based solar panels have been commercialized with a high apparent efficiency (>25%).[1] As the synthesis of low-dimensional silicon nanostructures develop, extensive studies have been placed on their booming novel applications originating from their attractive physical and chemical properties, which are different from those of bulk counterparts.[2–25]

One-dimensional (1D) nanowire has attracted much attention for solar energy conversion due to its unique chemical and physical properties.[16–25] The benefits of nanowires for solar energy conversion include: (1) With the confinement in two lateral dimensions and the unconstrained axial dimension, the nanowires can direct the migration of photogenerated carriers along the axial dimension and simultaneously allow for efficient charge transport. (2) By integrating multifunctional components within a nanowire, the created built-in potential favors the electron–hole pairs separation, drives charge transport towards two opposite directions and suppresses the bulk recombination of electrons and holes. (3) Geometry of nanowire arrays has a feature of low optical reflection and enhances the light trapping and absorption within nanowire arrays.

In this chapter, we have summarized the recent progress on the photocatalytic activity of 1D silicon nanowires (SiNWs) and their composites. The synthesis and characterization of the photocatalysts are also included, along with the discussion on the scientific and technological challenges in the silicon-based photocatalysis.

2. SiNWs for Photocatalysis

2.1 H-terminated SiNWs for photocatalysis

Silicon-based photovoltaic devices have been demonstrated with high efficiency for converting solar energy into electricity. However, the photocatalytic activity of silicon nanostructures are seldom reported due to the easy formation of silicon oxide barrier layer in the presence of air and the consequent blockage of transport of photogenerated charges onto the surface of silicon for desired redox reactions. The first example of SiNW photocatalysts was the H-terminated SiNWs, which deliver the catalytic activity in photodegradation of rhodamine B and oxidation of benzyl alcohol to benzoic acid.[26] H-terminated SiNWs were prepared by hydrofluoric acid (HF) treatment on SiNWs. The unexpected and unusual results indicated that H-terminated SiNWs exhibited higher photocatalytic activity than those of Pd-, Au-, Rh- or Ag-modified SiNWs and only less than that of Pt-modified SiNWs in the degradation of rhodamine B (Fig. 1(a)). Figure 1(b) presents the schematic of electron–hole generation and

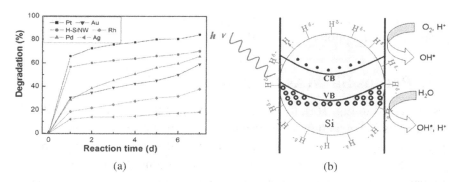

Fig. 1. (a) Photodegradation of rhodamine B under various SiNW catalysts at different times. The Pt-modified SiNWs had the best catalytic activity, followed by H-terminated, Pd-modified, Au-modified, Rh-modified and Ag-modified SiNWs. (b) Schematic of electron–hole generation in H-terminated SiNWs and their mechanism for the improved catalytic activity. Reprinted with permission from Ref. 26. © 2009 American Chemical Society.

separation and photocatalytic mechanism. Treated with HF, the surface of SiNWs is dominated by monohydride, dihydride and trihydride. The theoretical calculations indicate that the charges on the H atoms in three hydrides are electron-deficient. Therefore, the H atoms can serve as an electron sink and accelerate the separation of the photogenerated electrons and holes. The promoted photocatalytic activity of H-terminated SiNWs can be obtained.

The catalytic stability of H-terminated SiNWs was also prolonged. H-terminated SiNWAs still showed excellent catalytic activity after being immersed in HF solution for one week. It can be attributed to the modification of the surface of the SiNWs by fluorine ions. Fluorine, as the most electronegative element represents the most stable terminator, which can prevent the surface oxidation of silicon. However, the long-term catalytic stability of H-terminated SiNWs was not realized due to the decreased surface atomic ratio of F/Si for H-terminated SiNWs after immersion in solution for a long term.

Next, the high photocatalytic efficiency of H-terminated SiNWs synthesized by vapor–liquid–solid growth mechanism was reported under both visible and UV light irradiation.[27] Taking the degradation of rhodamine B as the demo reaction, the best photocatalytic activity was found on the H-terminated SiNWs decorated with Cu nanoparticles. The results indicated that the density of Cu nanoparticles seemed to be the dominant parameter in the photocatalytic efficiency.

2.2. Porous SiNWAs for photocatalysis

As presented in Fig. 1(a), the catalytic activity of the H-terminated SiNWs was unsatisfied, which may be attributed to the nature of the indirect bandgap of silicon. Recently, a novel single crystal silicon nanostructure — porous SiNWs with nanowalls of 3–5 nm, presents both electrical and optical activity.[28–32] Such unique properties of porous SiNWs originated from the strong quantum confinement effect and/or surface defects, in which the photogenerated electron–hole pairs can be efficiently separated and transferred onto the surface of SiNWs for the desired electrochemical reactions. The successful development of this multifunctional nanostructure may open new opportunities for a new generation of SiNWs-based photocatalysts with prominent visible-light absorption and high catalytic activity.[32]

Metal-assisted electroless chemical etching method is employed to synthesize porous SiNWs.[28–36] Generally, the metal-assisted chemical etching reactions can be classified into two types: one-step reaction and two-step reaction. Both approaches are found to effectively produce porous SiNWs. One-step chemical etching involves the immersion of the highly doped p-type silicon pieces (resistivity $< 0.005\ \Omega \cdot$ cm) in an etchant solution containing 0.01–0.04 M $AgNO_3$ and 5.0 M HF.[28] As-etched porous SiNWs were vertically oriented on silicon substrate (Fig. 2(a)). The selected area electron diffraction (SAED) (Fig. 2(b)) and high resolution transmission electron microscopy (HRTEM) (Fig. 2(c), 2(d)) indicated that porous SiNWs maintained their single crystalline structure as same as that of the starting silicon wafer.

Two-step chemical etching is employed to produce single crystalline porous SiNWs from n-type silicon wafers.[29–31] In the first step, the silver metal is deposited on silicon substrate in a solution composed of 0.005 M $AgNO_3$ and 4.8 M HF for 1 min. After thoroughly washing with water, the Ag-deposited silicon substrate is immersed into an etchant solution containing various concentrations of H_2O_2 and 4.8 M HF. Extensive studies show that the concentration of H_2O_2 and the doping level of silicon wafer are

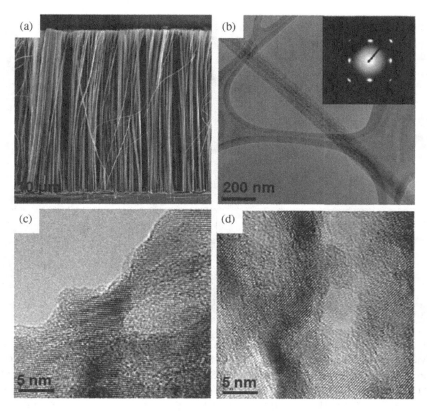

Fig. 2. Structural characterization of the porous SiNWs synthesized from highly doped p-type Si wafers (0.005 Ω·cm). (a) SEM image of the cross-section of the as-etched porous SiNWs. (b) TEM image of a typical porous SiNWs. Inset shows SAED pattern of porous SiNWs. (c,d) HRTEM images of the porous SiNWs. Adapted from Ref. 28. © 2009 American Chemical Society.

main factors to control morphology of As-synthesized SiNWs. When the lightly doped n-Si(100) substrate with a resistivity of larger than 0.3 Ω·cm is used, the non-porous SiNWs are generally produced. When the resistivity of n-Si(100) substrate is decreasing below 0.02 Ω·cm, the increase of the concentration of H_2O_2 results in a morphology evolution from pure solid nanowires (Fig. 3(a)) to rough-surfaced nanowires (Figs. 3(b,c) solid/porous core–shell nanowires (Figs. 3(d), 3(e)) and porous nanowires (Figs. 3(f)–3(h)). The porous SiNWs were vertically aligned on the silicon substrate (Fig. 3(i)). HRTEM studies indicate that as-grown SiNWs and porous

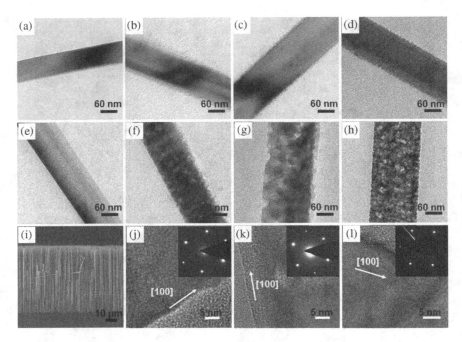

Fig. 3. Structural characterization of SiNWs from n-Si(100) with 0.008–0.02 Ω·cm resistivity in etchant solutions composed of 4.8 M HF and variable concentrations of H_2O_2 through a two-step reaction. TEM images of SiNWs (a) 0.1 M H_2O_2 for 30 min; (b) 0.1 M H_2O_2 for 60 min (c) 0.15 M H_2O_2 for 30 min; (d) 0.15 M H_2O_2 for 60 min; (e) 0.2 M H_2O_2 for 30 min; (f) 0.2 M H_2O_2 for 60 min; (g) 0.3 M H_2O_2 for 30 min and (h) 0.3 M H_2O_2 for 60 min. (i) SEM image of SiNWs in etchant solution with 0.3 M H_2O_2 for 60 min. HRTEM images of SiNWs (j) 0.1 M H_2O_2 for 30 min; (k) 0.2 M H_2O_2 for 30 min and (l) 0.3 M H_2O_2 for 60 min. Insets show SAED pattern of SiNWs. Adapted from Ref. 29. © 2009 American Chemical Society.

SiNWs (Figs. 3(j)–3(l)) retain the single crystalline structure of the starting silicon wafer.

Both n-type and p-type porous SiNWs show a broad visible emission near infrared range.[28,29] The visible emission, centered around 650 nm was observed. It may be generated from the deep quantum confinement in the porous structure. Electrical transport studies on individual nanowire, used in a simple back-gated device configuration on a silicon substrate suggest the electrically active behavior for the porous SiNWs. Hence, both optically and electrically active features of the porous SiNWs indicate that

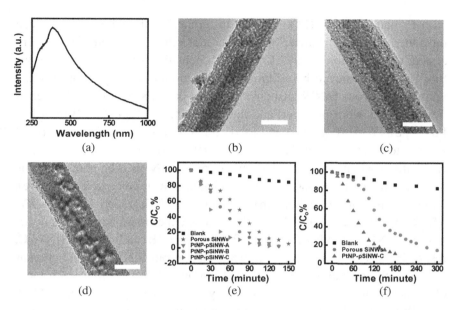

Fig. 4. Characterization and photoactivity of the porous SiNWs and PtNP loaded porous SiNWs. (a) Absorption spectrum of the porous SiNWs (0.1 mg/mL). (b) TEM image of the PtNP–pSiNW–A. (c) TEM image of the PtNP–pSiNW–B. (d) TEM image of the PtNP–pSiNW–C. (e) IC degradation catalyzed by the porous SiNWs and Pt loaded porous SiNWs. Black squares represent the IC only under the light irradiation. Blue stars, dark cyan down-triangles, magenta circles and dark yellow right-triangles represent the catalytic behavior of the porous SiNWs, PtNP–pSiNW-A, PtNP–pSiNW–B and PtNP-pSiNW-C, respectively. (f) 4-Nitrophenol degradation catalyzed by the porous SiNWs and PtNP loaded porous SiNWs. Black squares represent the 4-nitrophenol without photocatalysts. Red circles and blue up-triangles represent the catalytic behavior of the porous SiNWs and the PtNP–pSiNW–C, respectively. The scale bar is 100 nm. Reprinted with permission from Ref. 32. © 2010 Royal Chemical Society.

excitons generated within the porous SiNWs could be energetic enough to drive photoelectrochemical reactions.[32] The porous SiNWs show a much broader absorption (Fig. 4(a)) spanning over the entire spectral range from UV to visible and near IR. Therefore, porous SiNWs can be one of potential candidates for solar energy conversion.

Photocatalytic activity of porous SiNWs synthesized from highly doped n-Si (100) wafers (0.01–0.02 Ω·cm) has been investigated.[32] In order to enhance the photocatalytic activity, porous SiNWs were modified with Pt nanoparticles (PtNPs). The difference in their Fermi levels can introduce a Schottky barrier, which facilitates the separation of photogenerated electron–hole pairs. Controlled densities of PtNPs on porous SiNWs were achieved by adding various amounts of PtNPs into a fixed volume of porous SiNW suspension. Three catalyts with different densities of PtNPs were prepared and named as PtNP–pSiNW–A (Fig. 4(b)), PtNP–pSiNW–B (Fig. 4(c)) and PtNP–pSiNW–C (Fig. 4(d)), respectively.

Photodegradation of indigo carmine (IC) and 4-nitrophenol were carried out to evaluate the photoactivity of porous SiNWs. The photocatalytic activity of the four catalysts on the degradation of IC is compared in Fig. 4(e). After 60 min of irradiation, the percentages of degraded IC are 4.7%, 37.2%, 51.1%, 62.2% and 86.9% for control experiment without catalysts, porous SiNWs, PtNP-pSiNW-A, PtNP-pSiNW-B and PtNP-pSiNW-C, respectively. The results clearly indicate that porous SiNWs can function as effective photocatalysts and Pt-loaded porous SiNWs exhibit much more efficiency than the porous SiNWs only. The porous SiNWs and PtNP-pSiNW-C were also used to degrade 4-nitrophenol (Fig. 4(f)). 18.8 % and 66.0% of the 4-nitrophenol were degraded after 90 min of light irradiation for the porous SiNWs and the PtNP-pSiNW-C, respectively. The results further demonstrate the photocatalytic ability of the porous SiNWs and improved photoactivity of porous SiNWs by introducing PtNPs.

2.3. SiNW arrays for photocatalysis

Lately, the systematic studies on the photocatalytic activity of SiNWs were performed on the non-porous and porous silicon nanowire arrays (SiNWAs) from both *n*- and *p*-type silicon substrate with different doping levels.[37] By adopting the two-step chemical etching approach, the non-porous SiNWAs (defined as *p*-SiNWAs or *n*-SiNWAs) were obtianed from the Si substrate with a low doping concentration, while the porous SiNWAs (designated as *p*-pSiNWAs or *n*-pSiNWAs) were synthesized on the surface of the highly doped Si substrates. The absorption spectra of the four SiNWAs span across the entire UV and visible light (Fig. 5(a)), indicating the good light absorption of SiNWAs. The stronger absorption intensity of porous SiNWAs than that of the nonporous SiNWAs at the same substrate area can be attributed to their rough surface and large surface area of SiNWs. Treated with

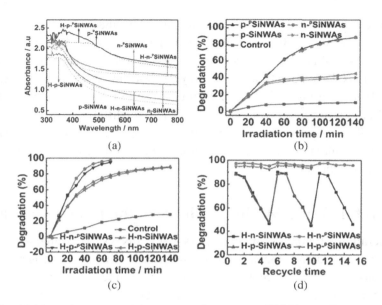

Fig. 5. (a) Absorption spectra of the porous and non-porous SiNWAs. (b) Catalytic activity of the porous and non-porous SiNWAs for photodegradation of methyl blue. (c) Catalytic activity of the H-terminated porous and H-terminated non-porous SiNWAs for photodegradation of methyl blue. (d) Photocatalytic stability of the H-terminated porous and H-terminated nonporous SiNWAs as a function of catalytic cycles. Reprinted with permission from Ref. 37. © 2011 Royal Chemical Society.

HF, the intensities of all absorption spectra decrease slightly due to the removal of silicon oxides, which may function as the anti-reflective layer.

The photocatalytic activity of SiNWAs was evaulated by the photodegradation of methyl red. As presented in Fig. 5(b), the four kinds of SiNWAs exhibited the photocatalytic activity under the visible-light irradiation. The catalytic efficiency of porous SiNWAs (p^{-p}SiNWAs and n^{-p}SiNWAs) were much higher than that of the non-porous SiNWAs (p^-SiNWAs and n^-SiNWAs). Of note, the catalytic activity of porous SiNWs was regardless of the type of silicon substrates. The higher catalytic activity of porous SiNWs is attributed to their larger surface areas and better light absorption. After removal of surface oxide by HF treatment, H-terminated SiNWs delivered promoted catalytic activity (Fig. 5(c)), which resulted from the more Si–H bond on the surface SiNWs for the enhanced charge separation and easy production of hydroxide radicals.

The catalytic stability of four kinds of H-terminated silicon nanowire arrays was evaluated by the cycling photodegradation reaction (Fig. 5(d)). Both p-PSiNWAs and n-PSiNWAs possessed robust catalytic stability. No obvious decayed activity for five cycles was observed. In contrast, the degradation activity of the non-porous SiNWAs quickly decreased to ~40% after five cycles. Although such a catalytic difference in the stability was attributed to more active sites after HF treatment and larger surface area, it was mainly due to the lack of fundamental understanding in their catalytic performance.

3. Metal Oxide Modified SiNWAs for Photocatalysis

Cooperation of other semiconductors with SiNWs is another effective approach to increase the separation and transport of the photogenerated electrons and holes. Recently, Cu_2O, a p-type semiconductor, was conjugated with n-type SiNWAs (Fig. 6(a)).[38] The composite showed an improved photocatalytic activity. The copper nanoparticles were firstly deposited on the surface of SiNWs by electroless chemical etching and the Cu_2O was formed by a subsequent thermal treatment in the presence of oxygen.

The photocatalytic activity of the SiNWAs and their composite was evaluated by the photodegradation of rhodamine B in the absence or presence of H_2O_2. As shown in Fig. 6(b), both SiNWAs and Cu_2O/SiNWAs exhibited the photocatalytic activity, compared to the negligible degradation of rhodamine B. The Cu_2O/SiNWA photocatalysts delivered a three times higher catalytic activity than that of SiNWAs. It was attributed to the formation of p/n heterojunctions between p-type Cu_2O nanoparticles and n-type SiNWs, which accelerated the separation of photogenerated electrons and holes effectively. After addition of H_2O_2, the catalytic activity of both SiNWAs and Cu_2O/SiNWAs was improved. H_2O_2 can react with excited electrons to form more hydroxide radicals and enhance the photocatalytic ability subsequently. Figure 6(c) shows the catalytic mechanism of the promoted activity of the composites. After direct contact between Cu_2O and SiNWs, an internal electric field is built. Under the light irradiation, the photogenerated electrons migrate to SiNWs and the photogenerated holes flow into Cu_2O nanoparticles. Thus, the recombination of the photogenerated electrons and holes is inhibited effectively, resulting in the improved photocatalytic activity of SiNWAs.

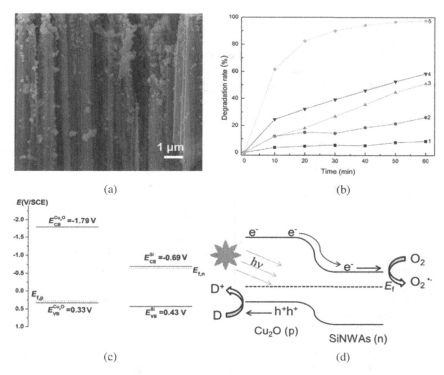

Fig. 6. (a) Scanning electron microscopy (SEM) image of Cu_2O/SiNWAs photocatalysts. (b) Photocatalytic degradation of RhB under different conditions: (1) SiNWAs; (2) Cu_2O/SiNWAs; (c) only H_2O_2; (4) SiNWAs+ H_2O_2; (5) Cu_2O/SiNWAs+ H_2O_2. The experiments were performed under visible-light irradiation with a power density of 100 mW/cm². The initial concentration of RhB was 10 mg/mL. For H_2O_2 addition, 0.5 mL of 30% H_2O_2 was used. (c) Schematic diagram of charge separation and transport between p-Cu_2O and n-SiNWAs. Reprinted with permission from Ref. 38. © 2014 Elsevier Publishing.

4. Stand Alone Pt/N–Si/Ag Nanowire Photocatalysts

Despite the good light absorption and beneficial charge conductivity of SiNWs, the investigation of SiNWs for photocatalysis is still limited due to their poor catalytic stability. The formation of silicon oxides significantly lowers the efficiency of charge transport. In contrast, the intensive studies on the improved catalytic stability of the silicon substrates based photo-electrochemical devices has been placed on the formation of the pinholes and cracks free protective layer, which not only isolates the silicon from the reactive environmentals but also improves the charge transportation due to

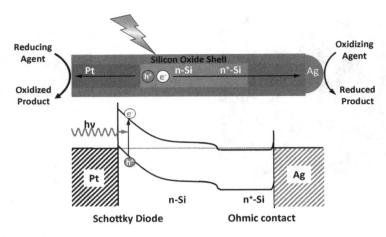

Fig. 7. Schematic illustration of a rationally designed metal-semiconductor metal-photoelectric nanodevice as a highly efficient photocatalyst. Here a Pt/n-Si/n^+-Si/Ag heterostructure is created in a single nanowire to integrate a nanoscale metal-semiconductor Schottky photodiode (Pt–Si) encased in a protective insulating shell with two exposed metal catalysts (Pt, Ag). Reprinted with permission from Ref. 48. Copyright 2010, American Chemical Society.

the formation of heterojunctions.[39–47] However, the investigation on the catalytic stability of the silicon nanostructures for photocatalysis is scant due to the difficulties in the formation of protective sheaths on the nanoscaled interface of nanosized silicon.

Recently, a unique 1D heterogeneous nanowire shows a highly efficient and stable photocatalytic activity by integrating multiple functional components.[48] As shown in Fig. 7, SiNWAs as the light absorbing antenna integrate with two noble metals Pd and Ag at two ends as the redox cocatalysts. Meanwhile, the two metals and silica shell as the protective shell to enwrap the silicon and its interfaces with Pt and Ag inside. The nanoscale Shottky photodiode is created between Pt and n-Si due to a large work function difference. A nearly Ohmic contact can be formed between the heavily doped Si (n^+-Si) and Ag due to a small work function difference. Under the light irradiation, the built-in potential (Pt/n-Si Shottky junction) as the driving force rapidly dissociates the photogenerated electron–hole pairs. The 1D morphology and protective shell ensure the efficient migration of the separated electrons and holes in the opposite directions and thus minimize the charge recombination. The protective components prevent the aggressive and deleterious side electrochemical reactions on silicon surface or their interfaces with Pt and Ag and ensure the photochemical stability of the catalytic system.

4.1 Pt/n-Si/n^+-Si/Ag nanowire photocatalysts and their catalytic activity[48]

1D Pt/n-Si/n^+-Si/Ag photodiode was synthesized via a dry etching method combining colloidal lithography with deep reactive ion etching (DRIE), electrodeposition of Pt and self-catalyzed Ag photodeposition, as illustrated in previous reports.[48] Figure 8(a) shows a typical TEM image of a free standing Pt/n-Si/n^+-Si nanowire, Fig. 8(b) shows Pt/n-Si/n^+-Si/Ag nanowires for self-catalyzed Ag deposition for short time and Fig. 8(c)

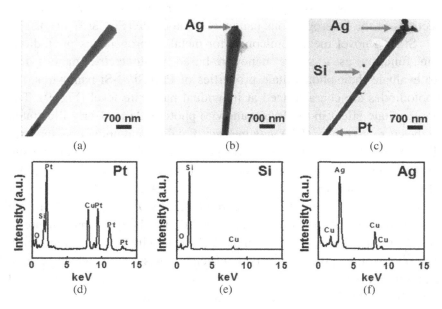

Fig. 8. Synthesis of Pt/n-Si/n-Si$^+$/Ag nanowire photodiodes. (a) Pt/n-Si/n-Si$^+$ nanowire before silver photoreduction. (b) Pt/n-Si/n-Si$^+$ nanowires dispersed in 1 mM AgNO$_3$ solution with 30 min light irradiation. (c) Pt/n-Si/n-Si$^+$/Ag nanowire diode obtained after 3 h light irradiation in 1 mM AgNO$_3$ solution. (e.g.) Energy dispersive X-ray spectrometry (EDX) of three distinct sections (Pt, Si and Ag) of a nanowire photodiode. Reprinted with permission from Ref. 48. Copyright 2010, American Chemical Society.

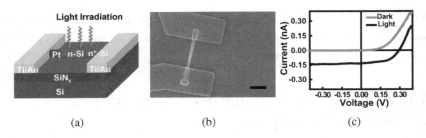

Fig. 9. (a) Schematic illustration of the fabricated device of individual Pt/Si nanowire photodiode. (b) SEM image of a Pt/Si nanowire device. Scale bar is 2 μm. (c) I–V data obtained in dark (light black) and under one-sun irradiation (dark black). Reprinted with permission from Ref. 48. Copyright 2010, American Chemical Society.

shows Pt/n-Si/n^+-Si/Ag nanowires for self-catalyzed Ag deposition for long time. All nanowires are encapsulated in a silicon oxide protective shell. Energy-dispersive X-ray (EDX) spectra confirmed three distinct sections of the heterogeneous nanowire photodiode (Fig. 8(d)–8(f)).

Such a novel metal-semiconductor-metal heterogeneous photodiode can function as a single nanowire-based photoelectric nanodevice. To evaluate their photovoltaic properties of Pt/n-Si/n^+-Si nanowires, the photodiodes are characterized at individual nanowire level (Fig. 9). The photovoltaic effect in the Pt/Si nanowire photodiode is clearly illustrated in the current-voltage (I-V) data measured in dark and under one-sun irradiation (Fig. 9(a), 9(b)). The I-V data (Fig. 9(c)) obtained under one-sun irradiation show clear photovoltaic effect with an open-circuit voltage V_{oc} of 0.30 V, a short-circuit current I_{sc} of 0.135 nA and a fill factor F_{fill} of 54%. The apparent short-circuit current density J_{sc} calculated from the projected active area is 11.2 mA/cm², which is much higher than the previous value of 3.5 mA/cm² for single Al/Si nanowire Schottky junction.[49] The photovoltaic power conversion efficiency η is 1.81%, which is more than thrice the 0.5% value reported for Si p–i–n junction nanowire[50] and the 0.46% value reported for single Al/Si nanowire.[49]

The 1D structure integrated a built-in heterogeneous photodiode and two redox cocatalysts improve the separation of photogenerated electron–hole pairs, transport of separated electrons and hole and efficient surface electrochemical reactions on two cocatalysts. The high catalytic performance is expected. Photovoltaic studies on the Pt/n-Si/n^+-Si/Ag

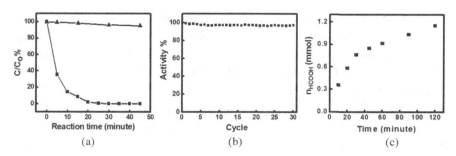

Fig. 10. Photocatalytic properties of Pt/n-Si/n^+-Si/Ag photodiode. (a) Photodegradation of IC with 1.2×10^8 Pt/n-Si/n^+-Si/Ag photodiodes dispersed in 10 mL of 100 μM IC aqueous solution. The triangles represent the photodegradation reaction of IC without photodiodes. The squares represent the photodegradation reaction of IC with Pt/n-Si/n^+-Si/Ag photodiodes. (b) Photocatalytic stability of the Pt/n-Si/n^+-Si/Ag photodiodes in the IC degradation reaction. (c) Plot of the total amount of formic acid with respect to the reaction time. In this reaction, 300 sccm CO_2 and 300 sccm H_2 were bubbled through 40 mL of 0.04 M Na_2CO_3 solution containing 1×10^8 Pt/n-Si/n^+-Si/Ag nanowire catalysts. Reprinted with permission from Ref. 48. Copyright 2010, American Chemical Society.

nanostructure have shown that the heterogeneous nanowires as a standalone photoelectric nanodevice can be employed for a wide range of either downhill or uphill reactions. Photocatalytic activity of the Pt/n-Si/n^+-Si/Ag nanowire photocatalysts exhibited high activity for the photodegradation of organic pollutants (Fig. 10(a), downhill reaction) and photocatalytic reduction of carbon dioxide (Fig. 10(c), uphill reaction). Meanwhile, the high electrochemical stability with minimal semiconductor degradation has been realized due to the presence of the protective SiO_2 on the sidewall and two metal cocatalysts at ends.

For photodegradation of IC, the reactions were irradiated under a xenon light and the change of IC concentration was monitored as a function of time. With the aid of Pt/n-Si/n^+-Si/Ag photocatalysts, 64.5% of IC molecules were degraded within 5 min (Fig. 10(a)). After 25 min irradiation, the IC molecules were degraded completely. In contrast, less than 2% of the IC was degraded with the same irradiation conditions in the absence of the nanowire photocatalysts. The calculated apparent quantum efficiency (QE) of the reaction is ~ 2.93% based on the initial 5 min of the reaction. It is also important to evaluate the photocatalytic stability of the photocatalysts. Under the same conditions, the photocatalytic activity of

Pt/n-Si/n^+-Si/Ag photodiodes remained essentially constant over 30 cycles (Fig. 10(b)), demonstrating the exceptional stability of Pt/n-Si/n^+-Si/Ag nanowire photodiodes in photocatalytic reactions. The features of the Pt/n-Si/n^+-Si/Ag photodiodes with the seamless integration of multiple functional components within a nanowire and the precise control of the material interface contribute to their high photocatalytic activity by boosting the efficient charge separation and transport toward the redox cocatalysts and improving the catalytic activity subsequently and their remarkable catalytic stability by integrating protective components and suppressing the deleterious side reactions on the surface of silicon or at the interfaces of metal/silicon.

The I-V testing (Fig. 9(c)) indicates that each Pt/n-Si/n^+-Si/Ag nanowire photodiode is a single nanoscaled photovoltaic cell with an open circuit voltage of 0.3 V. Thus, the Pt/n-Si/n^+-Si/Ag photodiodes can be used to catalyze thermodynamically uphill reactions. For example, synthesis of formic acid from CO_2 and H_2 requires an extra potential of 0.11 V. Fig. 10(c) shows the plot of the total amount of photoreduced formic acid vs. the reaction time. The plot shows a quick increase of formic acid at the beginning in the first 30 min followed by a slower linear increase. The calculated QE is about 19.9%.

4.2 Synthesis of Pt/Si/Ag nanowire photodiodes via wet chemical approach[51]

The synthetic process of Pt/n-Si/n^+-Si/Ag nanowire diodes, involving colloid nanolithography, dry etching, electrodeposition and self-catalytic photoreduction of Ag, is rather complicated, time-consuming and expensive. Ag-assisted electroless chemical etching may provide a simple, low-cost and scalable method to synthesize SiNWs. Typical TEM images of SiNW, SiNW with etched hole and Pt/Si heterojunction nanowire are shown in (Figs. 11(a)–11(c)). For formation of protective materials on the silicon and interface of Pt/Si and the end of SiNWAs, the oxides were formed by thermal annealing and Ag metal was deposited on the exposed silicon end via self-catalyzed photoreduction process as described above. A typical Pt/Si/Ag heterostructure is shown in Fig. 11(d). The Pt/n-Si forms a Schottky diode with a built-in potential due to large

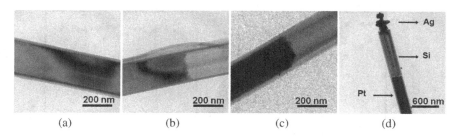

Fig. 11. Structural characterization of the nanowire heterostructures (a) TEM image of a wet etched SiNW after 900°C annealing in air. (b) TEM image of a SiNW with a nanohole. (c) Image of a heterogeneous Pt/Si nanowire. (d) TEM image of a Pt/Si/Ag nanowire. Reprinted with permission from Ref 51. Copyright 2010, Wiley Publishing.

work function difference (~1.5 eV), while a much less significant Schottky barrier also exists at Ag/n-Si interface because of their small work function difference (~0.2 eV). Therefore, this heterogeneous nanowire can work as a standalone photoelectric nanodevice for photocatalysis.

Fig. 12(a) shows the photodegradation rate of IC as a function of reaction time. About 16.1% of IC molecules were degraded in the first 15 min with the aid of the Pt/Si/Ag nanowires. The apparent QE of the photodegradation was estimated to be 0.58%, which is lower than that of the Pt/n-Si/n^+-Si/Ag photodiodes. This difference might be attributed to the small Schottky barrier between lightly doped silicon and silver, which may lower the efficiency of the electron transportation to silver catalysts compared to the Ohmic contact between highly doped Si and Ag in Pt/n-Si/n^+-Si/Ag nanowire photodiodes. After 210 min of light irradiation, 90.8% of IC was degraded. The high catalytic stability of the Pt/Si/Ag nanowire photocatalysts was also observed (Fig. 12(b)). Fig. 12(c) presents the degradation of 4-nitrophenol catalyzed by Pt/Si/Ag nanowire photocatalysts. Within the first 15 min of light irradiation, 10.3% of 4-nitrophenol molecules were degraded in the presence of the Pt/Si/Ag nanowire photodiodes. In contrast, only 1.7% of 4-nitrophenol molecules were degraded within the same period of time for the reaction in the absence of the nanowire photodiodes. The 4-nitrophenol molecules were completely degraded by the Pt/Si/Ag nanowire photodiodes after 660 min under light irradiation.

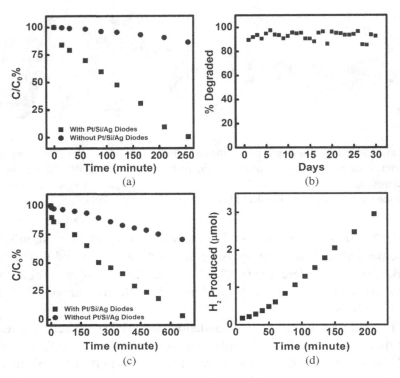

Fig. 12. Photocatalytic properties of the Pt/Si/Ag heterogeneous nanowires: (a) Photocatalytic degradation of IC with 2×10^9 Pt/Si/Ag photodiodes dispersed in 10 mL of 0.5 mM IC aqueous solution. (b) Photocatalytic stability of the Pt/Si/Ag photodiodes in the IC degradation reaction. The experiment is carried out with 2.0×10^7 diodes dispersed in 2 mL of 10 μM IC aqueous solution under natural sunlight irradiation. (c) Photodegradation of 4-nitrophenol with 8×10^8 Pt/Si/Ag photodiodes dispersed in 10 mL of 0.25 mM 4-nitrophenol aqueous solution. (d) Hydrogen production based on oxidation of formic acid catalyzed by Pt/Si/Ag diodes. Reprinted with permission from Ref. 51. Copyright 2010, Wiley Publishing.

Hydrogen generation from photo-decomposition of formic acid was shown in Fig. 12(d). The photocatalytic behavior of the photodiodes showed an initial activation stage followed by a stable hydrogen production rate at 2.64×10^{-4} μmol s^{-1}. The measured photon flux through the cell is 2.3×10^{17} photons s^{-1} and the calculated apparent average external quantum yield throughout the visible range for the Pt/Si/Ag photodiodes is 0.14%.

4.3 Enhanced photocatalytic activity of Pt/Si/Ag nanowires modified with Au@Ag core–shell nanorods[52]

The design of the heterogeneous Pt/Si/Ag nanowires allows efficient charge separation, transport and utilization to enable exceptional photocatalytic activity. However, the overall efficiency of the Pt/Si/Ag photodiodes can still be limited by relatively low optical absorption (<20%). Introducing plasmonic nanostructures on the surface of the photodiodes may concentrate the local light intensity, enhance the light absorption of the Pt/Si/Ag photocatalysts and improve the catalytic efficiency subsequently.[52] Notably, the presence of a thin insulating silicon oxide shell to electrically isolate the metal nanostructures can avoid the interface defects and/or charge transfer between semiconductor and metal, while allow unattenuated propagation of the optical field around the metal nanostructures and result in locally concentrated light.

Three Au or Au/Ag core–shell nanorods with different plasmonic peaks were used to modify on the surface of the Pt/Si/Ag nanowire photocatalysts through electrostatic force. Three catalysts are named as Au NRs-Diodes, Au/Ag NRs-A-Diodes and Au/Ag NRs-B-Diodes for Au nanorods, Au/Ag core/shell nanorods-A and Au/Ag core–shell nanorods-B, respectively. The absorption spectra of plasmonic nanorods are shown in Fig. 13. Photodegradation of nitrobenzene is performed to evaluate the photocatalytic activity of the photocatalysts and local surface plasmon resonance (LSPR) effect of the metal nanorods. Fig. 13(a) shows the concentration of the IC aqueous solution as a function of reaction time. The degraded percentages of nitrobenzene by pure photodiodes are 34.6% within 40 min of light irradiation. In contrast, the degradation rates were significantly increased when the Au NRs-Diodes, Au/Ag NRs-A-Diodes and Au/Ag NRs-B-Diodes were used. The apparent rates of photodegradation of nitrobenzene catalyzed by Au NRs-Diodes, Au/Ag NRs-A-Diodes and Au/Ag NRs-B-Diodes were 1.68 ± 0.08, 2.09 ± 0.09 and 3.18 ± 0.11 times faster than that catalyzed by Pt/Si/Ag nanowires alone. Due to the existence of silicon oxide shell, the accelerated catalytic activity in the presence of plasmonic nanorods has not originated from metal nanorods as the electron traps to aid electron–hole separation or as

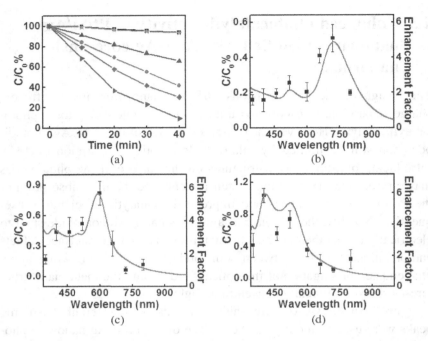

Fig. 13. (a) Photocalytic activity of four catalysts in nitrobenzene degradation reaction. Photocatalysts containing 2 mg of photodiodes were dispersed in 10 mL of 300 μM nitrobenzene aqueous solution. The blue up-triangles, red circles, olive diamonds and purple right-triangles represent the NB dye photodegradation catalyzed by photodiodes, Au NRs-Diodes, Au/Ag NRs-A-Diodes and Au/Ag NRs-B-Diodes, respectively. (b) Photoactivity enhancement factor (black square symbols) for Au NRs-Diodes as a function of excitation wavelength. Solid curve is the absorption spectrum of Au nanorods. (c) Photoactivity enhancement factor (black square symbols) for Au/Ag NRs-A-Diodes as a function of excitation wavelength. Solid curve is the absorption spectrum of Au/Ag core–shell nanorods. (d) Photoactivity enhancement factor (black square symbols) for Au/Ag NRs-B-Diodes as a function of excitation wavelength. Solid curve is the absorption spectrum of Au/Ag core–shell nanorods. Reprinted with permission from Ref. 52. Copyright 2011, American Chemical Society.

cocatalysts for surface electrochemical reactions. The enhanced catalytic activity of Pt/Si/Ag nanowires modified by plasmonic metal nanorods can be attributed to the effect of LSPR of metal nanostructures.

Spectral dependence of the photocatalytic activity was also investigated (Fig. 13(b)–13(d)). The photoactivity enhancement factor is defined as the values of IC degradation rates catalyzed by Pt/Si/Ag nanowires modified with metal nanorods to those catalyzed by Pt/Si/Ag photodiodes alone as a function of irradiation wavelength. The qualitative match between the enhancement factors at the selected irradiation wavelengths and the UV-vis absorption profiles of the metal nanostructures unambiguously demonstrates the metal nanorods plasmon resonance are responsible for the enhanced photoactivity of the Pt/Si/Ag photodiodes loaded with metal nanorods.

5. Conclusions

We have provided a concise overview of the recent research progress on 1D SiNWs and their composites for photocatalysis. These studies provide important foundation knowledge for further optimization of SiNW-based photocatalysts and development of novel heterogeneous SiNW composites for highly efficient photocatalysis. The current studies indicate that the photocatalytic performance of SiNW composites is still not competitive to many other photocatalysts due to easy formation of surface passive silicon oxide and the indirect bandgap of silicon. In the future, the research activity should be placed on exploring various heterogeneous structures for efficient charge separation and transport and improving the catalytic stability of SiNW composites. These studies can offer exciting potential in efficient utilization of solar energy.

Acknowledgement

We acknowledge the financial support from a NSFC Grant Nos. 21201138 and 21401148. This work was also partially funded by the Ministry of Science and Technology of China through a 973-program under Grant No. 2012CB619401 and supported by the Fundamental Research Funds for the Central Universities under Grant Nos. xjj2013102 and xjj2013043.

References

1. Retrieved from http://en.wikipedia.org/wiki/Solar_cell_efficiency#/media/File:PVeff%28rev150427%29a.jpg.
2. A. G. Cullis, L. T. Canham and P. D. J. Calcott, *J. Appl. Phys.* **82**, 1997, 909–965.

3. F. Priolo, T. Gregorkiewicz, M. Galli and T. F. Krauss, *Nat. Nanotechnol.* **9**, 2014, 19–32.
4. X. Y. Ye, S. Zou, K. X. Chen, J. J. Li, J. Huang, F. Cao, X. S. Wang, L. J. Zhang, X. F. Wang and M. R. Shen, *Adv. Funct. Mater.* **24**, 2014, 6708–6716.
5. R. Elnathan, M. Kwiat, F. Patolsky and N. H. Voelcker, *Nano Today* **9**, 2014, 172–196.
6. C. K. Chan, H. L. Peng, G. Liu, K. Mcllwrath, X. F. Zhang, R. A. Huggins and Y. Cui, *Nat. Nanotechnol.* **3**, 2008, 31–35.
7. M. Govoni, I. Mari and S. Ossicini, *Nature Photon.* **6**, 2012, 672–679.
8. C. Y. Liu, Z. C. Holman and U. R. L. Kortshagen, *Adv. Funct. Mater.* **20**, 2010, 2157–2164.
9. M. D. Kelzenberg, D. B. Turner-Evans, M. C. Putnam, S. W. Boettcher, R. M. Briggs, J. Y. Baek, N. S. Lewis and H. A. Atwater, *Energy Environ. Sci.* **4**, 2011, 866–871.
10. T. J. Kempa, B. Tian, D. R. Kim, J. Hu, X. Zheng and C. M. Lieber, *Nano Lett.* **8**, 2008, 3456–3460.
11. B. Tian, X. Zheng, T. J. Kempa, Y. Fang, N. Yu, G. Yu, J. Huang and C. M. Lieber, *Nature* **449**, 2007, 885–888.
12. X. Wang, K. Q. Peng, Y. Hu, F. Q. Zhang, B. Hu, L. Li, M. Wang, X. M. Meng and S. T. Lee, *Nano Lett.* **14**, 2014, 18–23.
13. S. K. Choi, U. Kang, S. Lee, D. Ham, S. M. Ji and H. Park, *Adv. Energy Mater.* **4**, 2014, 1301614.
14. Y. F. Li, M. C. Li, D. D. Song, H. Liu, B. Jiang, F. Bai and L. H. Chu, *Nano Energy* **11**, 2015, 756–764.
15. Y. L. Li, Q. Chen, D. Y. He and J. S. Li, *Nano Energy*, **7**, 2015, 10–24.
16. D. Liu, L. L. Li, Y. Gao, C. M. Wang, J. Jiang and Y. J. Xiong, *Angew. Chem. Int. Edit.* **54**, 2015, 2980–2985.
17. B. Tian, T. J. Kempa and C. M. Lieber, *Chem. Soc. Rev.* **38**, 2009, 16–24.
18. Y. Qu and X. F. Duan, *Chem. Soc. Rev.* **42**, 2013, 2568–2580.
19. E. L. Warren, H. A. Atwater and N. S. Lewis, *J. Phys. Chem. C* **118**, 2014, 747–759.
20. H. Zhou, Y. Qu and X. F. Duan, *Energy Environ. Sci.* **5**, 2012, 6732–6743.
21. Y. Qu and X. F. Duan, *J. Mater. Chem.* **22**, 2012, 16171–16181.
22. A. I. Hochbaum and P. Yang, *Chem. Rev.* **110**, 2010, 527–546.
23. K. Q. Peng and S. T. Lee, *Adv. Mater.* **23**, 2011, 198–215.
24. M. Law, J. Goldberger and P. D. Yang, *Ann. Rev. Mater. Res.* **34**, 2004, 83–122.
25. K. Q. Peng, X. Wang, L. Li, Y. Hu and S. T. Lee, *Nano Today* **8**, 2013, 75–97.

26. M. W. Shao, L. Cheng, X. H. Zhang, D. D. D. Ma and S. T. Lee, *J. Am. Chem. Soc.* **131**, 2009, 17738–17739.
27. N. Megouda, Y. Cofininier, S. Szunerits, T. Hadjersi, O. Elkechai and R. Boukherroub, *Chem. Comm.* **47**, 2011, 991–993.
28. A. I. Hochbaum, D. Gargas, Y. J. Hwang and P. D. Yang, *Nano Lett.* **9**, 2009, 3550–3554.
29. Y. Qu, L. Liao, Y. J. Li, H. Zhang, Y. Huang and X. F. Duan, *Nano Lett.* **9**, 2009, 4539–4543.
30. X. Zhong, Y. Qu, Y. C. Lin, L. Liao and X. F. Duan, *ACS Appl. Mater. Interf.* **3**, 2011, 261–270.
31. Y. Qu, H. Zhou and X. F. Duan, *Nanoscale* **3**, 2011, 4060–4068.
32. Y. Qu, X. Zhong, Y. J. Li, L. Liao, Y. Huang and X. F. Duan, *J. Mater. Chem.* **20**, 2010, 3590–3594.
33. H. H. Chen, R. J. Zou, N. Wang, Y. G. Sun, Q. W. Tian, J. H. Wu, Z. G. Chen and J. Q. Hu, *J. Mater. Chem.* **21**, 2011, 801–805.
34. X. L. Wang and W. Q. Han, *ACS Appl. Mater. Interf.* **2**, 2010, 3709–3713.
35. J. Kim, H. Han, Y. H. Kim, S. H. Choi, J. C. Kim and W. Lee, *ACS Nano* **5**, 2011, 3222–3229.
36. C. Chiappini, X. W. Liu, J. R. Fakhoury and M. Ferrari, *Adv. Funct. Mater.* **20**, 2010, 2231–2239.
37. F. Y Wang, Q. D. Yang, G. Xu, N. Y. Lei, Y. K. Tsang, N. B. Wong and J. C. Ho, *Nanoscale* **3**, 2011, 3269–3276.
38. C. T. Yang, J. L. Wang, L. R. Mei and X. Y. Wang, *J. Mater. Sci. Technol.* **30**, 2014, 1124–1129.
39. B. Seger, A. B. Laursen, P. C. K. Vesborg, T. Pedersen, O. Hansen, S. Dahl and I. Chorkendorff, *Angew. Chem. Int. Edit.* **51**, 2012, 9128–9131.
40. K. Sun, X. Pang, S. Shen, X. Qian, J. S. Cheung and D. Wang, *Nano Lett.* **13**, 2013, 2064–2072.
41. K. Sun, N. Park, Z. Sun, J. Zhou, J. Wang, X. Pang, S. Shen, S. Y. Noh, Y. Jing, S. Jin, P. K. L. Yu and D. Wang, *Energy Environ. Sci.* **5**, 2012, 7872–7877.
42. S. Mubeen, J. Lee, N. Singh, M. Moskovits and E. W. Mcfarland, *Energy Environ. Sci.* **6**, 2013, 1633–1639.
43. B. Seger, T. Pedersen, A. B. Laursen, P. C. K. Vesborg, O. Hansen and I. Chorkendorff, *J. Am. Chem. Soc.* **135**, 2013, 1057–1064.
44. Y. W. Chen, J. D. Prange, S. Dühnen, Y. Park, M. Gunji, C. E. D. Chidsey and P. C. McIntyre, *Nature Mater.* **10**, 2011, 539–544.
45. D. V. Esposito, I. Levin, T. P. Moffat and A. A. Talin, *Nature Mater.* **12**, 2013, 562–568.

46. M. J. Kenney, M. Gong, Y. Li, J. Z. Wu, J. Feng, M. Lanza and H. J. Dai, *Science* **342**, 2013, 836–840.
47. S. Hu, M. R. Shaner, J. A. Beardslee, M. Lichterman, B. S. Brunschwig and N. S. Lewis, *Science* **344**, 2014, 1005–1009.
48. Y. Qu, L. Liao, R. Cheng, Y. Wang, Y. C. Lin, Y. Huang and X. F. Duan, *Nano Lett.* **10**, 2010, 1941–1949.
49. M. D. Kelzenberg, D. B. Turner-Evans, B. M. Kayes, M. A. Filler, M. C. Putnam, N. S. Lewis and H. A. Atwater, *Nano Lett.* **8**, 2008, 710–714.
50. T. J. Kempa, B. Tian, D. R. Kim, J. Hu, X. Zheng and C. M. Lieber, *Nano Lett.* **8**, 2008, 3456–3460.
51. Y. Qu, T. Xue, X. Zhong, Y. C. Lin, L. Liao, J. Choi and X. Duan, *Adv. Funct. Mater.* **20**, 2010, 3005–3011.
52. Y. Qu, R. Chen, Q. Su and X. Duan, *J. Am. Chem. Soc.* **133**, 2011, 16730–16733.

Chapter 4

The Applications of Graphene-based Nanocomposites in the Field of Photocatalytic Selective Organic Transformations

Min-Quan Yang, Siqi Liu, Bo Weng and Yi-Jun Xu[*]

State Key Laboratory of Photocatalysis on Energy and Environment
College of Chemistry, Fuzhou University
Fuzhou, 350002, PRC
College of Chemistry,
New Campus, Fuzhou University, Fuzhou, 350108, PRC
[*]Corresponding author. [*]yjxu@fzu.edu.cn

1. Introduction

Since the pioneering work in 1972 that water could be photolyzed electrochemically at an illuminated Pt–TiO_2 electrode to yield stoichiometric quantities of H_2 and O_2,[1] heterogeneous photocatalysis has received intense attention and been widely utilized for environmental remediation and solar energy conversion.[2–11] The basic mechanism of semiconductor-based photocatalysis is that under the irradiation of sunight or an illuminated light source, a photocatalyst can absorb photons with energy that matches or exceeds the bandgap of the semiconductor.[4,5,12] The electrons in the valence band of the photocatalyst can be photoexcited to the conduction band, leaving a positive hole in the valence band. The photogenerated electron–hole pairs have important role for driving photocatalytic reactions.[5–6] However, the photoinduced electron–hole pairs in the excited states are unstable and can easily recombine at or near the surface of photocatalysts, dissipating the input energy as heat and/or light,[11,13] which is harmful for the photocatalytic application and significantly limits the efficiency of semiconductor-based photocatalysis.

In order to improve the photocatalytic activity of semiconductor-based photocatalysts, a variety of strategies have been developed. Thereinto, considerable attention has been devoted to integrating carbonaceous nanomaterials with semiconductor photocatalysts due to their unique and controllable structural and electrical properties, which has proven to be beneficial for enhancing the photocatalytic activities.[6,14–19] In recent decade, ignited by the advent of graphene (GR) in 2004, which has proven that the atomic monolayer crystal could be obtained on top of non-crystalline substrates,[20] there has been a significantly increasing research interest in designing novel GR-based materials for a variety of technological applications.[21–29] In particular, the two-dimensional (2D) sheet structure and high specific surface of GR make it an ideal platform for assembling semiconductor components and the excellent electronic properties of GR endow it with the prominent capability to accept/transport photogenerated charge carriers, thereby improving the efficiency of semiconductor-based artificial photocatalytic processes.[11, 23, 28–30]

To date, various GR-based composite photocatalysts with enhanced photocatalytic performance have been designed and widely applied in a myriad of fields.[11,23,28–32] An overview of literatures reveals that the recent progress in the applications of GR-based composite photocatalysts are mainly focused on "non-selective" degradation of pollutants in vapour and liquid phase, photoinactivation of bacteria and water splitting to H_2 and O_2. Research works on utilizing GR-based semiconductor nanocomposites for photocatalytic "selective" redox reaction are relatively limited. This is mainly due to the fact that heterogeneous photocatalysis as an advanced oxidation process is always considered to be highly unselective, especially in water.[12,33] Nevertheless, recent progress has proven that photocatalytic selective processes can be achieved through the selection of appropriate photocatalysts (e.g. semiconductor and organocatalyst) and control of the reaction conditions.[7,12,33–47] In comparison with conventional approaches for organic synthesis, photocatalysis exhibits significant advantages due to its possibility to employ milder conditions, achieve shorter reaction sequences and minimize undesirable side reactions, which provides an alternative environmentally friendly and cost-effective method for selective synthesis of fine chemicals.[12,48–50]

In this chapter, we highlight and summarize recent progress on GR-based photocatalysts for selective organic synthesis processes, including reduction of CO_2 to renewable fuels, reduction of nitroaromatic compounds to amino compounds, oxidation of alcohols to aldehydes and acids, epoxidation of alkenes, oxidation of primary C–H bonds, hydroxylation of phenol and oxidation of tertiary amines. The different roles of GR in these GR-based composite photocatalysts (e.g. electron reservoir, macromelecular photosensitizer), the effect of graphene oxide (GO, the precursor of GR) in GO-organic species photocatalysts and GO itself as a photocatalyst have been discussed. In addition, the strategies of strengthening the interfacial contact of GR and semiconductor, constructing multi-components GR-based composite photocatalysts and optimizing the atomic charge carrier transfer pathway across the interface between GR and semiconductor for further improving the photoactivity of GR-semiconductor composites has also been presented. It is hoped that the present chapter can provide useful information to broaden the applications of GR-based composites for solar energy conversion, especially in the field of photocatalytic selective organic transformations.

2. Photocatalytic Selective Reduction of Nitroaromatics

In recent years, the nitroaromatics resulting from the production of pesticides, herbicides, insecticides and synthetic dyes, have aroused great public concern because of their high pollution of water.[51–55] However, the products obtained from catalytic reduction of nitroaromatics (e.g. aromatic amines), are potent intermediates in industrial synthesis of biologically active compounds, pharmaceuticals, rubber chemicals, photographic and agricultural chemicals.[51,56,57] Therefore, the selective reduction of nitroaromatics to the corresponding amines has aroused a great deal of attention in both academic as well as technological fields.

Recently, Chang and co-workers have prepared a series of GR–ZnO–Au nanocomposites (GR–ZnO–Au NCs) via a simple hydrothermal method and tested their photoactivity toward selective reduction of nitrobenzene (NB) to aniline.[47] As shown in Fig. (1a), it can be seen that in the presence of methanol as hole scavenger, the sample of GR–ZnO–Au displays the highest photocatalytic activity, which can be ascribed to the critical role of GR in the composites. The introduction of GR and Au nanorods into the matrix of ZnO effectively improves the separation and transfer efficiency of electron–hole pairs (Fig. 1(b)), thus leading to the enhanced photoactivity of GR–ZnO–Au. In addition, the surface assisted laser desorption/ionization mass spectrometry has been applied to detect the intermediates (nitrosobenzene and phenylhydroxylamine) and major product (aniline) of NB through photocatalytic reactions. The result

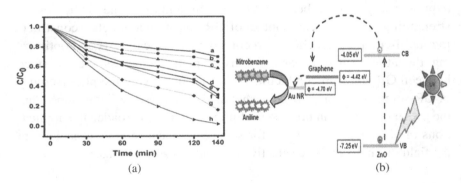

Fig. 1. (a) Reduction rates of NB (5 mM) in the absence of any photocatalyst a) and in the presence of GR (0.5 mg mL^{-1}) b), Au nanorods (0.5 mM) c), ZnO nanospheres (18 mM) d), P25 (18 mM) e), GR–Au nanorods (4 mg mL^{-1}) (f), GR–ZnO nanospheres (4 mg mL^{-1}) g) and GR–ZnO–Au nanocomposite (4 mg mL^{-1}) (h). C_0 and C are the concentrations of NB at reaction times 0 and t, respectively; (b) schematic diagram showing a plausible mechanism involved in the transfer of electrons from the different energy levels of the GR–ZnO–Au nanocomposite for the photoreduction of NB (note: CB and VB correspond to the conduction band and valence band edges of ZnO. The symbol Φ corresponds to the work function of GR and Au). Reprinted with permission from Ref. 47. Copyright 2013, American Chemical Society.

reveals that the reduction of NB to aniline is through a favored pathway of photoreduction NB to nitrosobenzene and then to aniline directly.

Our group has fabricated the CdS nanowires (CdS NWs)-GR nanosheets composite with 1D-2D structure[58] and CdS nanosphere-GR (CdS NSP-GR) nanohybrids with quasi core–shell structure[59] via electronic self-assembly method, followed by GO reduction through a hydrothermal treatment. The (CdS NWs)-GR and (CdS NSP-GR) nanocomposites have been tested for photocatalytic selective reduction of a series of nitroaromatics to their corresponding amines (Scheme 1). The results demonstrate that the (CdS NWs)-GR and (CdS NSP-GR) nanocomposites both exhibit considerable visible-light photoactivity. The photoactivities of (CdS NWs)-GR and (CdS NSP-GR) are also much higher than CdS NWs and CdS NSPs due to the fact that the GR sheets can increase the adsorptivity of reactants and promote charge carriers separation and transportation. In addition, the photocorrosion of CdS can be efficiently inhibited in the reaction system of adding ammonium formate ($HCOONH_4$) as hole scavenger and under N_2 atmosphere, demonstrating that CdS–GR can perform as stable visible-light-driven photocatalyst toward selective photoreduction of nitroaromatics through the proper control of reaction conditions.

On the basis of the work of CdS NWs-GR, our group has further constructed a ternary nanohybrid of reduced GO–CdS nanowires-TiO_2 (CTG) featuring a large 2D flat structure via a simple surface charge promoted self-assembly method.[60] Compared to the curly reduced GO–CdS NWs (CG) synthesized by a similar approach, the negatively charged TiO_2 nanoparticles (NPs) decorated on both the surfaces of CdS nanowires (CdS NWs) and RGO sheets can prevent the RGO sheets from being curly or aggregated through an electrostatic repulsion, thereby forming the large

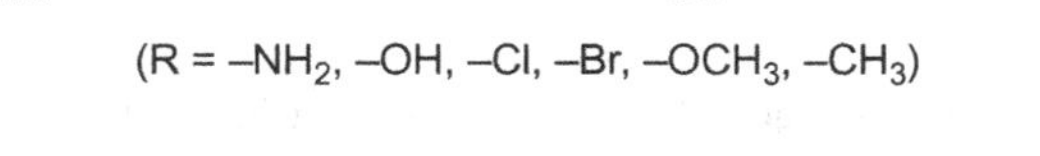

$$R\text{-}C_6H_4\text{-}NO_2 \xrightarrow[N_2,\ HCOONH_4]{h\nu,\ photocatalyst} R\text{-}C_6H_4\text{-}NH_2$$

(R = $-NH_2$, $-OH$, $-Cl$, $-Br$, $-OCH_3$, $-CH_3$)

Scheme 1. Photocatalytic reduction of nitro compounds to amino compounds in water with the addition of ammonium formate ($HCOONH_4$) as hole scavenger under N_2 atmosphere.

2D flat structure of CTG, as shown in Figs. 2(a), (b). This large 2D flat structure of CTG provides a large surface area and more efficient contact between light and RGO sheets, leads to an increased optical absorption of visible-light, facilitates the migration of the photogenerated charge carriers. In addition, the TiO_2 NPs on CdS NWs are also able to further boost the transfer and prolong the lifetime of charge carriers in the ternary CTG system due to the suitable energy band match of TiO_2 to CdS (Figs. 2(c), (d) and thus enhance the overall photocatalytic performance of CTG.

Very recently, Liu *et al.* have reported a facile layer-by-layer self-assembly approach (Fig. 3(a)) for the construction of well-defined GR

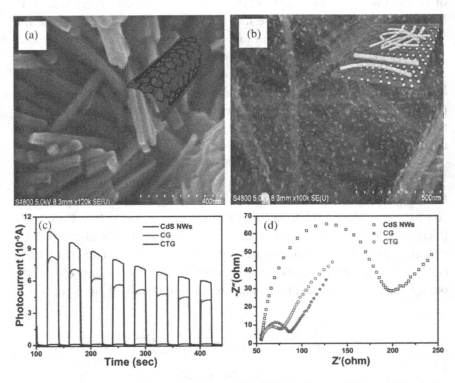

Fig. 2. (a) Typical SEM images of CG and (b) CTG (the insets of (a) and (b) are the corresponding schematic modes); (c) transient photocurrent responses and (d) electrochemical impedance spectroscopy (EIS) Nyquist plots of CdS NWs, CG and CTG. Reprinted with permission from Ref. 60. Copyright 2014, Royal Society of Chemistry.

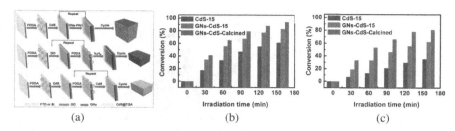

Fig. 3. (a) Schematic illustration for layer-by-layer self-assembly of GNs–CdS QDs, pure GNs and pure CdS QDs multilayered films; photocatalytic reduction of substituted nitroaromatic compounds over CdS QDs film (15 cycles), as-assembled GNs–CdS QDs composite film (15 cycles) and calcined GNs–CdS QDs composite films (15 cycles) under visible-light irradiation ($\lambda > 420$ nm), with the addition of ammonium formate as quencher for photogenerated holes and N_2 purge under ambient conditions: (b) 4–nitroaniline, (c) 4–nitrophenol. Reprinted with permission from Ref. 61. Copyright 2014, American Chemical Society.

nanosheets–CdS quantum dots (GNs–CdS QDs) multilayered films.[61] The composite films are composed of tailor-made negatively charged CdS QDs and positively charged GNs-poly(allylamine hydrochloride), which are judiciously stacked in an alternating manner based on pronounced electrostatic interaction. By this means, large area, smooth and uniform hybrid films have been fabricated. Additionally, the architecture, photo-electrochemical and photocatalytic properties of the GNs–CdS QDs films can be easily tuned by simple control of deposition cycles. The GNs–CdS QDs multilayered films display promising photocatalytic performances towards photoreduction of nitroaromatics to corresponding amines under visible-light irradiation, as reflected by two representative examples in Figs. 3(b), (c). The photoactivity enhancement can be attributed to the judicious integration of CdS QDs with GNs in an alternative stacking manner, which well utilizes the 2D nanoarchitecture of GR and maximizes the charge separation and transport in GNs–CdS QDs composite films. This work is believed to boost further interest for fabricating uniform semiconductor/GNs hybrid films with controllable film thickness and architecture.

Apart from the typical semiconductor CdS, our group has prepared the In_2S_3–GR nanocomposites via the surface charge modified method,

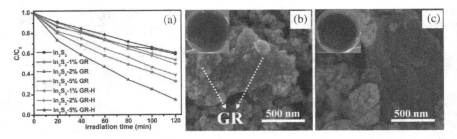

Fig. 4. (a) Photocatalytic selective reduction of 4-nitroaniline to p-phenylenediamine over the samples of In_2S_3, In_2S_3–GR and In_2S_3–GR–H nanocomposites with different weight additions of GR; typical SEM images of (b) In_2S_3–2%GR and (c) In_2S_3–2% GR–H, the insets of (a) and (b) are the corresponding photographs of In_2S_3–2%GR and In_2S_3–2%GR–H obtained after the hydrothermal treatment process. Reprinted with permission from Ref. 62. Copyright 2013, American Chemical Society.

which have also been applied to photocatalytic selective reduction of nitroaromatic compounds to amino organics.[62] As shown in Fig. 4(a), the obtained In_2S_3–GR nanocomposites display higher visible-light photoactivity than bare In_2S_3 nanoparticles. In addition, it is worth noting that the photoactivities of In_2S_3–GR are also higher than In_2S_3–GR–H samples that are obtained from the simple "hard" integration of GR nanosheets with solid In_2S_3 nanoparticles without modification of surface charge. The enhanced photocatalytic performance of In_2S_3–GR is mainly ascribed to the more efficient interfacial contact between In_2S_3 and the GNs than In_2S_3–GR–H (Figs. 4(b),(c)), which would be favorable to utilize the excellent electron conductivity of GR to transfer the photogenerated charge carriers and lengthen the lifetime of charge carriers more effectively. This work highlights the significant influence of interfacial contact between semiconductors and the GR nanosheets on determining the photoactivity of GR-semiconductor nanocomposites. In order to construct effective GR-semiconductor nanocomposites, the efficient utilization of the electron conductivity of GR by controlling interfacial interaction between GR and the semiconductor would be an important factor.

3. Photocatalytic Selective Reduction of CO_2

As energy and environmental issues caused by the use of fossil fuels have become serious, utilization of renewable energy resources has been increasingly critical. CO_2, a major anthropogenic greenhouse gas, has caused growing concern due to its continuous rise in atmospheric concentration and therefore various strategies have been developed to reduce it.[31,63,64] In this regard, the photocatalytic conversion of CO_2 into valuable solar fuels such as methane or methanol becomes a promising approach to both address the future energy supply demand and mitigate the problem of global warming.[64] Thus far, some GR-based nanocomposites have been successfully utilized for photocatalytic reduction of CO_2 into renewable fuels.

Zou et al. have fabricated robust titania–GR hollow spheres that consist of non-stoichiometric $Ti_{0.91}O_2$ nanosheets and GNs via a layer-by-layer assembly technique with polymer beads as sacrificial template and followed by a microwave irradiation technique to simultaneously remove the template and reduce GO into GR (Fig. 5(a)).[65] Toward the photocatalytic conversion of CO_2 into renewable fuels (CO and CH_4) in the presence of water vapor, the total conversion of CO_2 over $GR–Ti_{0.91}O_2$ hollow

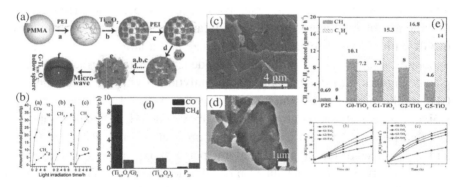

Fig. 5. (a) Schematic illustration of procedure for preparing the LBL-assembled multilayer-coated spheres consisting of titania nanosheets and GO nanosheets, followed by microwave reduction of GO into GR; (b) photocatalytic CH_4 (dots) and CO (squares) evolution amounts for $(GR–Ti_{0.91}O_2)_5$ hollow spheres a), $(Ti_{0.91}O_2)_5$ hollow spheres b) and P25 c); comparison of the average product formation rates d); (c) FE-SEM and (d) TEM image of $G2–TiO_2$ nanocomposite; (e) comparison of photocatalytic activity of samples $GRx–TiO_2$ 2 (x = 0,1,2,5; x is the weight ratio of G in the obtained samples) and P25, the molar ratio of C_2H_6 to CH_4 increases from 0.71 (for $G0–TiO_2$), 2.09 ($GR1–TiO_2$), 2.10 ($GR2–TiO_2$), to 3.04 ($GR5–TiO2_2$) a); photocatalytic CH_4 b) and C_2H_6 c) evolution amounts for samples $GRx–TiO_2$ (x = 0, 1, 2, 5). Reprinted with permission from Refs. 63 and 65. Copyright 2013, 2012, John Wiley & Sons, Inc.

spheres is five-times higher than blank-$Ti_{0.91}O_2$ hollow spheres and nine-times higher than commercial P25, as shown in Fig. 5(b). The superior photoactivity of GR–$Ti_{0.91}O_2$ hollow spheres can be attributed to that: (i) the ultrathin nature of $Ti_{0.91}O_2$ nanosheets is favorable for the rapid transportation of photoelectrons onto the surface to participate in the photoreduction; (ii) the sufficiently compact stacking of ultrathin $Ti_{0.91}O_2$ nanosheets with GNs leads to the spatial separation of the charge carriers, thus improving the efficiency of the photocatalytic process; (iii) the hollow structure allows a more efficient, permeable absorption and scattering of light, thereby contributing to the photoactivity enhancement. In a subsequent work, Zou *et al.* have reported another TiO_2–GR hybrid nanocomposite, which was prepared by an *in situ* reduction-hydrolysis technique in a binary ethylenediamine (En)–H_2O solvent.[63] The reduction of GO into GR by En and the formation of TiO_2 nanoparticles loaded onto GR through chemical bonds (Ti-O-C bond) are achieved simultaneously. Due to the reducing role of En, abundant Ti^{3+} is formed on the surface of the TiO_2 in the obtained G–TiO_2 nanohybrids. The typical SEM and TEM images of GR2 (2% GR)–TiO_2 nanohybrid in Figs. 5(c) and (d) indicate that the as-obtained nanocomposite maintains the 2D sheet-like structure of GR with a several-micrometer lateral size and a large number of TiO_2 nanoparticles are spread uniformly and densely on the surface of graphene. Figures 5(e) displays the photocatalytic activity for reduction of CO_2 in the presence of water vapor over the GR–TiO_2 samples, from which it is seen that in contrast with the above GR–$Ti_{0.91}O_2$ nanocomposites that produce CO and CH_4, the main products over G–TiO_2 are CH_4 and C_2H_6. The different photocatalytic performance is ascribed to the synergistic effect of the introduction of GR and the surface Ti^{3+} sites. The existence in abundance of Ti^{3+} sites plays a vital role for the coupling of photo-formed ·CH_3 radicals into C_2H_6. On the other hand, the electron-rich GR may help stabilize the ·CH_3 species via π-conjugation between the unpaired electron of the radical and aromatic regions of GR, which also raises the opportunity of formation of C_2H_6.

In addition to using GO as the precursor of GR for the preparation of GR–TiO_2 composite photocatalysts, Hersam and co-workers have reported the utilization of solvent exfoliated GR (SEG) to prepare SEG-P25 nanocomposites toward photocatalytic reduction of CO_2 to methane (CH_4).[66]

For comparison, the solvent-reduced GR oxide (SRGO)-P25 nanocomposites have also been prepared. As shown in Fig. 6(a), the SEG-P25 nanocomposites exhibit a significantly larger enhancement in CO_2 photoreduction under both UV and visible-light. In contrast, no improvement in the photoactivity is observed for SRGO-P25 films under UV light illumination and the photoactivity enhancement of the optimal SRGO-P25 nanocomposite is also limited under visible illumination. The differences in photoactivity of SEG-P25 and SRGO-P25 nanocomposites are due to the different defect densities and sheet resistance between them (Figs. 6(b), (c)). The SEG sheet with lower defect density and sheet resistance has superior electronic coupling to TiO_2, which enables photoexcited energetic electrons to diffuse farther from the GR-P25 interface, thus decreasing the likelihood of their recombination with holes on the TiO_2. This work

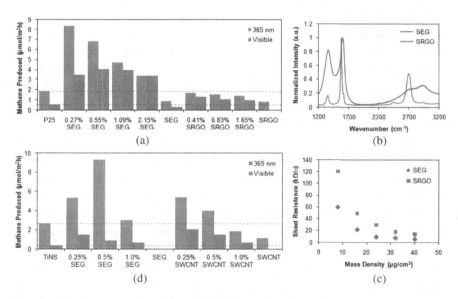

Fig. 6. (a) Photocatalytic activity of SEG-P25 and SRGO-P25 nanocomposites for CO_2 photoreduction under ultraviolet and visible illumination; (b) intensity-normalized Raman spectra of SEG and SRGO films annealed at 400°C for 30 min in air; (c) sheet resistances of SEG and SRGO thin films formed via vacuum filtration as a function of mass density and (d) photocatalytic activity of SEG–TiNS and SWCNT–TiNS nanocomposites for CO_2 photoreduction to CH_4 under ultraviolet and visible irradiation illumination. Reprinted with permission from Refs. 66 and 67. Copyright 2011, 2012, American Chemical Society.

demonstrates that the use of defect-free GR would be a straightforward strategy to exploit the outstanding electronic properties of GR and thus further improving the photocatalytic performance of GR-based photocatalysts toward target applications. In another work, Hersam *et al.* have proven that the dimensionality of carbon nanomaterial is also a key factor in determining the interfacial charge transfer and hence the photoactivity of carbon–titania nanosheet composite photocatalysts.[67] They have synthesized the non-covalently bound SEG–TiNS (titania nanosheets) composites and the single-walled carbon nanotubes (SWCNTs)–TiNS nanocomposites. The resulting 2D–SEG–TiNS and 1D–2D SWCNT–TiNS composites have been tested for the photocatalytic reduction of CO_2 in the presence of water under UV and visible-light irradiation. Both hybrid systems demonstrate an enhanced photocatalytic rate of formation of methane in comparison with blank-TiNS, as shown in Fig. 6(d). More importantly, because of the 2D nature of SEG in 2D–SEG–TiNS, they possess more intimate interfacial contact and stronger optoelectronic coupling with TiNS, which result in an enhanced photoactivity for CO_2 conversion as compared to 1D-2D SWCNT–TiNS under UV light illumination. Alternatively, under visible-light irradiation, SWCNT–TiNS display higher photoactivity for CO_2 reduction which is ascribed to the sensitization effect of SWCNTs on TiNS to absorb longer wavelength light.

Apart from semiconductor-GR composites, GO was also found to be a promising photocatalyst for the catalytic conversion of CO_2 to methanol (MeOH). Hsu *et al.* have conducted a systematic investigation of photocatalytic CO_2 reduction on various GO samples synthesized under different conditions.[68] They found that the GO-3, obtained by the modified Hummers' method in the presence of excess $KMnO_4$ and excess H_3PO_4 to raise the level of oxidation under the protection of GO basal plane, exhibits the highest photocatalytic efficiency among the studied samples, as shown in Fig. 7(a). The photocatalytic conversion rate of CO_2 to methanol on GO-3 is six-fold higher than the pure TiO_2. The possible photocatalytic mechanism has been proposed. As illustrated in Fig. 7(b), in the modified GO, the oxygenated functional groups provide a 2D network of sp^2 and sp^3 bonded atoms, leading to the presence of a finite bandgap depending on the isolated sp^2 domains. During the photocatalytic reduction process, the modified GO with surplus oxygenated components is photoexcited to

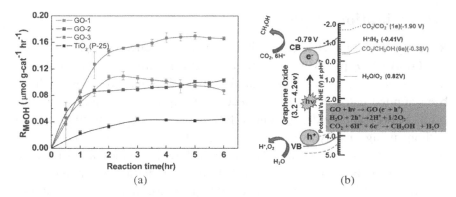

Fig. 7. (a) Photocatalytic methanol formation (R_{MeOH}) on different GO samples (GO–1, GO–2, GO–3) and TiO_2 using a simulated solar-light source; (b) schematic illustration of the photocatalytic CO_2 reduction mechanism on GO. Reprinted with permission from Ref. 68. Copyright 2013, Royal Society of Chemistry.

generate electron–hole pairs (e^-–h^+), which then migrate to the GO surface to react with absorbed reactants.

4. Photocatalytic Selective Oxidation Reactions

Selective oxidation processes are a type of the most studied reactions both in industrial and laboratory synthesis, which play an important role in the production of a wide range of chemicals.[12,50] The products and derivatives (e.g. aldehydes, ketone and hydroquinone) obtained from the oxidation processes are widely utilized in the fragrance, confectionary and pharmaceutical industries.[69–71] Typically, the classical industrial oxidation processes need harsh reaction conditions and strong oxidants, such as high temperature and pressure, mineral acids, chromates, permanganate.[12, 58, 72–73] Whereas, the technology of photocatalysis is driven by clean solar energy and performed under mild reaction conditions (i.e. room temperature, ambient atmosphere) with benign oxidants, such as O_2 and H_2O_2, which is moderate and environment friendly and can be cogitated as a possible promising approach to avoid the shortcomings of traditional industrial oxidation processes.[12,48–50] Regarding the photocatalytic selective oxidation processes, GR-based composite photocatalysts have also been studied and been proven to be effective.

4.1 Oxidation of phenol

Recently, Das and co-workers have prepared reduced graphene oxide (RGO)–Ag_3VO_4 photocatalysts via a one-pot *in situ* photochemical synthesis route.[74] The as-synthesized RGO–Ag_3VO_4 nanocomposites have been applied to photocatalytic selective hydroxylation of phenol to catechol (CAT) and hydroquinone (HQ) under visible-light illumination without using any hydroxylating agents. The conversion of phenol and selectivity towards CAT and HQ over different samples are shown in Table 1. It can be seen that under visible-light irradiation of 2 h, the introduction of GR into the matrix of Ag_3VO_4 gives rise to a prominent photoactivity enhancement and changes the selectivity distribution of the products accordingly. The conversion of phenol and selectivity for CAT improves with increase of the RGO content. This is mainly due to the photocatalytic hydroxylation process over the RGO–Ag_3VO_4 nanocomposites in aqueous medium, the RGO sheets are able to accept electrons easily, forming the superoxide radical anion ($O_2^{\cdot-}$) that might be subsequently transformed to a hydroxyl radical ($\cdot OH$), thus facilitating the hydroxylation

Table 1: Catalytic activity and product selectivity over various RGO–Ag_3VO_4 photocatalysts[74]

Catalyst	Conversion (%)	Selectivity (%)	
		CAT	HQ
No catalyst	0	—	—
Graphite	0	—	—
GO	0	—	—
RGO	0	—	—
Ag_3VO_4	18.82	—	90
1RGO–Ag_3VO_4	36.47	32.9	67.1
2RGO–Ag_3VO_4	66	40.25	59.75
4RGO–Ag_3VO_4	100	80.23	19.77
8RGO–Ag_3VO_4	100	89.38	10.62
4GN–Ag_3VO_4	70.24	50.49	49.51

Reaction conditions: [catalyst] = 2 g L^{-1}, [phenol] = 20 mgL^{-1}, VIS light irradiation time = 2 h.

reaction. This work provides a green and effective process for the synthesis of catechol from hydroxylation of phenol in water under visible-light irradiation.

4.2 Oxidation of alcohols and alkenes

Our group has carried out a series of relative studies about the utilization of GR-based composite photocatalysts for selective oxidation of alcohols and alkenes.[34,35,58,75–79] For instance, we have prepared GR–TiO_2 nanocomposites with intimate interfacial contact via a facile wet-chemistry approach during which GO was used as the precursors of GR.[77] The GR–TiO_2 nanocomposites have been used for the photocatalytic selective oxidation of various benzylic alcohols and allylic alcohols to their corresponding aldehydes with high selectivity under visible-light irradiation, as shown in Table 2. In addition, compared with the CNT–TiO_2 composites prepared by the similar procedure, the GR–TiO_2 composites exhibit

Table 2: Selective oxidation of a range of alcohols over the TiO_2–5% GR photocatalyst under the visible-light irradiation ($\lambda > 400$ nm) for 20 h[77]

Entry	Substrate	Product	Conversion (%)	Yield (%)	Selectivity (%)
1	C6H5-CH2OH	C6H5-CHO	62	62	100
2	4-CH3-C6H4-CH2OH	4-CH3-C6H4-CHO	70	70	100
3	4-CH3O-C6H4-CH2OH	4-CH3O-C6H4-CHO	80	80	100
4	4-O2N-C6H4-CH2OH	4-O2N-C6H4-CHO	74	73	99
5	4-Cl-C6H4-CH2OH	4-Cl-C6H4-CHO	45	43	96
6	4-F-C6H4-CH2OH	4-F-C6H4-CHO	84	76	91
7	cinnamyl alcohol	cinnamaldehyde	50	46	92
8	prenol	prenal	41	37	90

higher photoactivity. This is mainly due to that GO with abundant surface oxygen-containing functional groups displays superior and easily accessible "structure-directing" role, leading to a more intimate interfacial contact between GR and TiO_2 than that of CNT–TiO_2 (Figs. 8(a) and (b)). The intimate complexation of TiO_2 and GR is beneficial for facilitating the separation and transfer of the photogenerated charge carriers, thus enhancing the photocatalytic activity of GR–TiO_2.

In the following work of our group, a series of GR–CdS nanocomposites with different weight addition ratios of GR have been synthesized by a facile one-step *in situ* solvothermal strategy.[58] Figure 8(d) displays the typical SEM image of CdS–GR, from which it can be seen that the CdS nanoparticles with small size are intimately spreaded on the surface of GR with large layered structure. As for the sample of blank-CdS, spherical CdS particles with the average diameter around 200 nm are obtained in the absence of GR (Fig. 8(c)). The result demonstrates that GO (the precursor of GR) has a significant influence on the morphology of CdS. Toward photocatalytic selective oxidation of various alcohols to their corresponding aldehydes under ambient conditions, CdS–5% GR exhibits optimally enhanced photocatalytic performance as compared to blank-CdS. This photoactivity enhancement can be attributed to the effective hybridization of GR with CdS, which significantly influences the properties of the CdS–GR samples, including morphology, optical and electronic nature as compared with blank -CdS.

The above two representative works demonstrate the possibility of utilizing GR-based composite photocatalysts for selective oxidation of

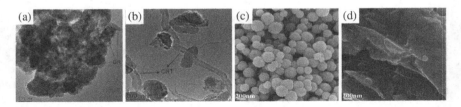

Fig. 8. TEM images of (a) TiO_2–5% GR and (b) TiO_2–5% CNT; SEM images of the as-prepared samples of (c) blank-CdS, (d) CdS–5% GR. Reprinted with permission. Reprinted with permission from Refs. 77,58. Copyright 2011, 2011, American Chemical Society.

alcohols, which enriches the potential applications related to GR-based composite photocatalysts. In addition, the works also reveal the 2D structural advantage of GO on changing the structure and morphology of the samples, which is beneficial for engineering the interfacial contact between GR derived from GO precursor and semiconductors to fabricate efficient GR-based semiconductor nanocomposites with high photocatalytic performance.

On the basis of the CdS–GR composite, we have further constructed ternary CdS–GR–TiO$_2$ nanocomposites using an *in situ* growth strategy with TiO$_2$ nanoparticles uniformly carpeted on the surface of the CdS–GR.[76] The introduction of TiO$_2$ into the CdS–GR substrate has no impact on the morphology and porosity properties of the samples, whereas it improves the photoactivity of CdS–GR–TiO$_2$ toward selective oxidation of benzylic alcohols and allylic alcohols as compared to the binary CdS–GR counterparts, as reflected in Fig. 9. The improved photocatalytic activity is ascribed to the combined interaction of the longer lifetime of photogenerated electron–hole pairs, faster interfacial charge transfer rate and larger surface area. This work demonstrates that the careful design of GR-based composites by coupling GR with multiple semiconductor compounds is favorable to the spatial transfer pf photogenerated charge carriers, which would have the potential to improve their capacity for photocatalytic processes significantly.

In the previous work of Hersam *et al.*[66] they have shown that the quality of the precursor of GR has significantly influence on the photoactivity of the prepared GR-semiconductor composite photocatalysts. The solvent

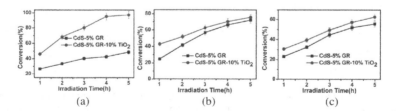

Fig. 9. Time-online photocatalytic selective oxidation of alcohols to aldehydes over the optimal CdS-5% GR-10% TiO$_2$ and CdS-5% GR photocatalysts under visible-light irradiation s: (a) benzyl alcohol, (b) cinnamyl alcohol and (c) 3-methyl-2-buten-1-ol. Reprinted with permission from Ref. 76. Copyright 2011, American Chemical Society.

exfoliated GR (SEG) with lower defect density is beneficial for improving the photoactivity of the P25 semiconductor. Ignited by this work, our group has fabricated a series of GR (SEG and GR)–TiO$_2$ nanocomposites following a facile "soft" chemistry approach using GO and SEG as the precursor of GR, respectively.[78] For comparison, the nanocomposites of SEG-P25 and GR-P25 have also been prepared by a simple "hard" integration method. The photoactivity tests (Fig. 10(a)) for selective oxidation of different benzylic alcohols and allylic alcohols under visible-light irradiation demonstrate that the as-synthesized SEG-TiO$_2$ and SEG-P25 display higher photoactivities as compared to their counterparts of GR–TiO$_2$ and GR-P25. Moreover, the SEG-TiO$_2$ and GR–TiO$_2$ obtained by the "soft" integration method display higher photoactivities than SEG-P25 and GR-P25 prepared by the "hard" integration method. The photoactivity results indicate that (i) the SEG, with lower numbers of defects (Fig. 10(b)), can make more efficient use of the electron conductivity of GR, by which the separate and transfer of photoexcited charge carriers can be improved more effectively, as supported by the photoluminescence (PL) spectra analysis in Fig. 10(c); (ii) the rational design and fabrication of GR-semiconductor nanocomposites play a key role in determining their photocatalytic performance toward specific applications.

Very recently, our group has reported a simple, conceptual and general approach to significantly improve the photoactivity of GR-semiconductor

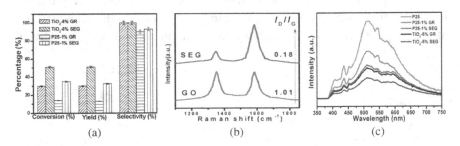

Fig. 10. (a) the comparison of photocatalytic selective oxidation of benzyl alcohol over the nanocomposites of TiO$_2$–5% SEG, TiO$_2$–5% GR, P25–1% SEG and P25–1% GR under visible-light irradiation of 4 h at room temperature; (b) Raman spectra of GO and solvent-exfoliated GR (SEG); (c) photoluminescence (PL) spectra of the samples TiO$_2$–5% SEG, TiO$_2$–5% GR, P25–1% SEG, P25–1% GR and bare P25 nanoparticles. Reprinted with permission from Ref. 78. Copyright 2012, Royal Society of Chemistry.

nanocomposites via introducing a tiny amount of metal ions (Ca^{2+}, Cr^{3+}, Mn^{2+}, Fe^{2+}, Co^{2+}, Ni^{2+}, Cu^{2+} and Zn^{2+}) into the interfacial layer matrix between the GR sheets and semiconductor CdS,[80] as illustrated in Fig. 11(a). This simple strategy can not only significantly improve the visible-light-driven photoactivity of GR–CdS nanocomposites for targeting selective photoredox reaction, including aerobic oxidation of alcohol (Fig. 11(b)) and anaerobic reduction of nitro compound (Fig. 11(c)), but also drive a balance between the positive effect of GR on retarding the recombination

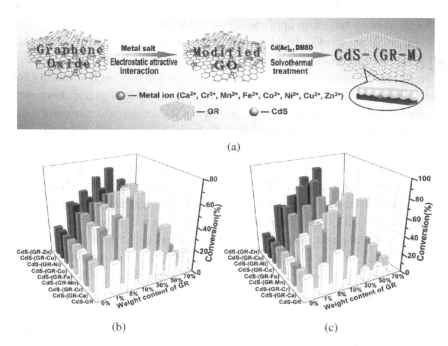

Fig. 11. Flowchart illustrating the fabrication of CdS–(GR–M) (M = Ca^{2+}, Cr^{3+}, Mn^{2+}, Fe^{2+}, Co^{2+}, Ni^{2+}, Cu^{2+} and Zn^{2+}) nanocomposites in which metal ions are introduced to the interfacial layer matrix between GR and semiconductor CdS (a); photocatalytic performance of blank-CdS, CdS–GR and CdS–(GR–M) (M = Ca^{2+}, Cr^{3+}, Mn^{2+}, Fe^{2+}, Co^{2+}, Ni^{2+}, Cu^{2+} and Zn^{2+}) nanocomposites with different weight addition ratios of GR for (b) photocatalytic selective oxidation of benzyl alcohol under visible-light ($\lambda > 420$ nm) for 2 h and (c) selective reduction of 4-nitroaniline under visible-light irradiation ($\lambda > 420$ nm) for 80 min. Reprinted with permission from Ref. 80. Copyright 2014, American Chemical Society.

of electron–hole pairs photogenerated from semiconductor and the negative "shielding effect" of GR resulting from the high weight addition of GR. The optimal weight addition ratio of GR is remarkably increased from 5% to 10% or 30% for the series of GR–CdS and CdS–(GR–M) photocatalysts. This work highlights that the optimization of interfacial composition between GR and semiconductors (e.g. introducing metal ions or other heteroatoms as interfacial mediator) could offer a conceptually new strategy to boost the lifetime and transfer efficiency of photogenerated charge carriers across the interface between GR and semiconductors. As a result, the more efficient and smarter GR-semiconductor photocatalysts would be achieved towards various targeting applications in solar energy conversion.

In the above works, the photocatalytic selective oxidation reactions are all performed in the organic solvent benzotrifluoride (BTF). The choice of solvent BTF is because of its inertness to oxidation and high solubility for molecular oxygen. Recently, driven by the typical tenet of green chemistry, we have also investigated the application of GR-semiconductor composite photocatalysts to selective oxidation of alcohols in water,[75] which is safe, low-cost and green. We have synthesized the CdS–GR (RGO, SEG)[75] and TiO_2–GR (RGO, SEG)[81] nanocomposites with different weight addition ratios of GR. The photoactivity test results for oxidation of benzylic alcohols demonstrate that, similar to the case in the organic solvent of BTF, CdS–GR and TiO_2–GR composites are able to serve as selective visible-light photocatalysts toward oxidation of benzylic alcohols in the green solvent water, during which the products are mainly composed of aldehydes and acids. As exemplified in Table 3, it can be seen that the introduction of an appropriate amount of GR can effectively enhance the photoactivity of the CdS–GR and the selectivity distribution of the products is changed accordingly. The high weight addition of GR in CdS–GR[75] and TiO_2–GR[81] nanocomposites will lead to a decrease of selectivity for benzaldehyde, with the main byproduct of benzoic acid. Additionally, it can also be found that the CdS–SEG[75] and TiO_2–SEG[81] nanocomposites both exhibit obviously enhanced photocatalytic activity compared to the CdS–RGO and TiO_2–RGO counterparts due to the fact that SEG has a better electron conductivity than RGO, which is beneficial for separating photogenerated charge carriers and inhibiting the recombination of electrons and holes, thus improving the photoactivity. The results would

Table 3: Photocatalytic selective oxidation of various benzylic alcohols to aldehydes and acids over blank-CdS, CdS–5% RGO and CdS–5% SEG nanocomposites in water under visible-light irradiation for 4 h[75]

Entry	R	blank-CdS Conv.(%)	Yield(%) Aldehyde	Acid	CdS–5% RGO Conv.(%)	Yield(%) Aldehyde	Acid	CdS–5% SEG Conv.(%)	Yield(%) Aldehyde	Acid
1	p-H	30	13	17	35	25	10	67	47	20
2	p-methyl	24	17	7	28	21	7	55	40	15
3	p-methoxyl	36	17	19	50	32	18	68	45	23
4	p-nitro	29	6	23	37	8	29	54	12	42
5	p-fluoro	30	6	24	41	10	31	50	15	35
6	p-chloro	23	8	15	32	12	20	72	19	53

enrich the applications of GR-semiconductor nanocomposites in photocatalytic selective organic transformations in water.

Besides the typical semiconductors of CdS and TiO_2, ZnS has also been investigated to couple with GR for photocatalytic selective oxidation of various alcohols under visible-light irradiation.[79] The series of ZnS–GR nanocomposites with different weight ratios of GR were synthesized by a facile two-step wet chemistry process, during which the nanosized ZnS particles were tightly anchored on the 2D platform of GR with an intimate interfacial contact. The UV-vis diffuse spectra of ZnS–GR nanocomposites shows that the introduction of GR into the substrate of ZnS is able to narrow the bandgap of ZnS to some degree, but it is still not narrow enough to the visible-light region. However, the photoactivity tests over the ZnS–GR nanocomposites demonstrate that the as-prepared ZnS–GR exhibits obvious visible-light photoactivity while the blank-ZnS displays negligible photoactivity toward aerobic selective oxidation of alcohols and epoxidation of alkenes under visible-light illumination (Fig. 12). Moreover, the control experiments reveal that the visible-light photocatalytic reaction mechanism of ZnS–GR is essentially different from the case under UV light irradiation. There are no holes involved in the photocatalytic oxidation of alcohols or alkenes under visible-light. Based on the joint characterizations and in terms of a structure-photoactivity correlation analysis, it is proposed that the GR in the ZnS–GR nanocomposites is able to be photoexcited from the ground state GR to the excited state GR* and then injects electrons into the conduction band of ZnS, which serve as an

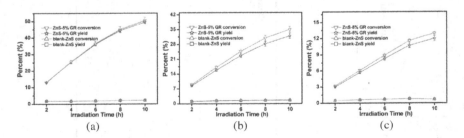

Fig. 12. Time-online profiles of photocatalytic selective oxidation of organic compounds over blank-ZnS and ZnS-5%GR under the irradiation of visible light; (a) benzyl alcohol; (b) 3-methylbut-2-en-1-ol; (c) styrene. Reprinted with permission from Ref. 79. Copyright 2012, American Chemical Society.

organic dye-like macromolecular "photosensitizer" instead of an electron reservoir. This novel photocatalytic mechanism is distinctly different from previous studies on GR-semiconductor photocatalysts, for which GR is claimed to behave as an electron reservoir to capture and shuttle the electrons photogenerated from the semiconductor. This study is hoped to provide a new train of thought on designing GR-based composite photocatalysts for targeting applications and, fundamentally, promote in-depth thinking on the microscopic charge carrier transfer pathway connected to the interface between the GR and semiconductor.

4.3 Oxidation of primary C–H bonds

Selective oxidation of saturated sp^3 for fine chemicals synthesis is of crucial importance for the sustainable exploitation of available feedstocks. In a recent study, we have reported a viable strategy to synthesize ternary GR–CdS–TiO_2 composites with an intimate spatial integration and sheet-like structure by assembling two co-catalysts, GR and TiO_2, into the semiconductor CdS matrix with specific morphology as a visible-light harvester.[82] The ternary GR–CdS–TiO_2 composites are able to serve as a highly selective visible-light-driven photocatalyst for oxidation of saturated primary C–H bonds using benign oxygen as oxidant under ambient conditions. Table 4 shows the results of photocatalytic selective oxidation of toluene and other substituted toluenes to corresponding aldehydes over the blank-CdS, optimal 5% GR–CdS and 5%GR–CdS–10%TiO_2 photocatalysts under visible-light irradiation. It can be seen that the optimal 5%GR–CdS composite displays much higher photoactivity than the blank-CdS photocatalyst and the introduction of the second co-catalyst TiO_2 further improves the photoactivity of CdS catalyst toward C–H activation under identical reaction conditions. The comparative photoelectrochemical and photoluminescence (PL) spectra analysis of the blank-CdS, optimal 5% GR–CdS and 5%GR–CdS–10%TiO_2 indicate that the introduction of dual co-catalysts into the CdS matrix optimizes and improves the separation and transfer of charge carriers over the light harvester CdS under visible-light irradiation. This work further enriches the applications of GR-semiconductor nanocomposites in photocatalytic selective organic transformations and it also demonstrates that the rational

Table 4: Photocatalytic selective oxidation of toluene and other substituted toluenes to corresponding aldehydes over blank-CdS, 5%GR–CdS and 5%GR-CdS–10%TiO$_2$ under visible- light irradiation of 10 h[82]

Catalyst	Substrate	Product	Conv. (%)	Sel. (%)
blank-CdS			33	100
5%GR–CdS			53	99
5%GR-CdS–10%TiO$_2$			69	98
blank-CdS			27	100
5%GR–CdS			52	99
5%GR-CdS–10%TiO$_2$	Cl	Cl	66	99
blank-CdS			30	100
5%GR–CdS			54	99
5%GR-CdS–10%TiO$_2$	F	F	68	99
blank-CdS			29	100
5%GR–CdS			51	99
5%GR-CdS–10%TiO$_2$	NO$_2$	NO$_2$	65	98
blank-CdS			36	100
5%GR–CdS			57	100
5%GR-CdS–10%TiO$_2$	O–	O–	71	99
blank-CdS			39	100
5%GR–CdS			58	98
5%GR-CdS–10%TiO$_2$			73	98

optimization of the chemical compositions and synergetic interaction of GR and semiconductors is a promising strategy to design more efficient semiconductor-based photocatalyst.

4.4 Tertiary amines oxidation and Mannich reaction

Recently, the incorporation of visible-light-active organocatalyst with GO has been demonstrated to improve the photoactivity of organic synthesis reactions. Tan and co-workers have reported the oxidative C–H functionalization of tertiary amines by the combination of GO and Rose Bengal (RB) under green light irradiation.[83] Table 5 displays the photoactivities for a-cyanation of tertiary amine over different catalysts under similar reaction conditions. In comparison with RB alone, the addition of GO into the reaction system obviously enhances the conversion and yield of tertiary amines. The controlled experiments show that no desired product is observed when the reaction is conducted in the dark and only trace amount of product is detected when the reaction is performed without RB, demonstrating that light and photoredox catalyst are both essential for this reaction. In addition, GO can be recycled without the loss of its activity and does not catalyze this reaction by itself alone, indicating that GO behaves as a co-catalyst in the composite catalyst system. Taking the optimal 5 mol% RB and 50 wt.% GO as an example, the performance of photocatalytic a-cyanation of a series of cyclic tertiary amines have also

Table 5: a-Cyanation of tertiary amine in the presence of RB and GO[a83]

Entry	Carbon catalyst	Yield (%)
1	none	60
2	50 wt.% GO[c]	97
3	50 wt.% graphite	65
4	50 wt.% activated carbon	56
5	200 wt.% GO	96
6[d]	50 wt.% GO	94 (63)
7[e]	50 wt.% GO	10
8[f]	50 wt.% GO	0
9[g]	50 wt.% GO	94

[a]Reaction was performed using 0.05 mmol of 1 and 0.125 mmol of TMSCN in 0.5 mL of CH_3CN.
[b]Isolated yield.
[c]GO was prepared according to the reported method.[16]
[d]$Ru(bpy)_3Cl_2$ was used instead of RB. Number in bracket is the yield without GO.
[e]Reaction was performed without RB.
[f]Reaction was performed in dark.
[g]Recycled GO was used.

been evaluated. Similar observations are made; the presence of GO increases the reaction rate and also improves the overall yield of the reactions. This work provides the example of using GO as a co-catalyst to facilitate the synthesis of small molecular organic compounds under visible-light irradiation.

In another work by the group, the authors have constructed hybrid graphene oxide (GO)-poly(3–hexylthio-phene-2,5–diyl) (P3HT) and RGO (reduced graphene oxide) –P3HT nanocomposites assembled via π-π interaction between the non-polar regions of GO/RGO and the molecular layers of P3HT.[84] The obtained GO–P3HT and RGO–P3HT hybrids display broad-band absorption. The joint characterizations of photothermal desorption spectroscopy, time-resolved photoluminescence studies and transient absorption spectra demonstrate that the semiconducting conjugated polymer of P3HT could be photoexcited to generate charge pairs, which then are injected into GO/RGO across the interface of the GO/RGO–P3HT within a very short time. The photocatalytic activities of the GO/RGO–P3HT for the oxidative Mannich reaction are tested under mild conditions. As displayed in Table 6, the GO–P3HT shows the highest yield (93%) incomparison with P3HT (65%) and RGO–P3HT (60%).

Table 6: Oxidative Mannich Reaction between N-Aryl–etrahedroisoquinoline and Acetone[a,84]

Entry	Catalyst	Amount (wt.%)	Yield (%)[b]
1	No catalyst		47
2	GO	2.5	44
3	r–GO–P3HT	2.5	64
4	GO–P3HT	2.5	93
5	P3HT	2.5	74
6	P3HT	0.5	65

[a]The reaction was performed using 0.1 mmol of 1 in the mixed solvent of 0.5 mL of acetone and 0.5 mL of $CHCl_3$.
[b]Isolated yield.

The enhanced photocatalytic performance of GO–P3HT is proposed to be due to the following. Under the ambient sunlight irradiation, the charge carriers are generated from the excitation of P3HT. The tertiary amine is oxidized by the positive hole on the HOMO of P3HT *via* single electron transfer to form the radical cation. At the same time, the excited electron is injected from the LUMO of P3HT into GO films, which is then used to activate molecular oxygen to form the dioxygen radical anion. The intermediate dioxygen radical anion is stabilized by the aromatic scaffold in GO and the oxygenated groups in the GO surface may undergo electrostatic interaction with the amine radical cations, thus facilitating the reaction with the dioxygen radical anion to form iminium. Compared to GO–P3HT, the smaller population of oxygenated species on the RGO–P3HT scaffold is suggested to be the possible reason for its lower catalytic activity.

Conclusion

In summary, this chapter focused on the applications of GR-based composite photocatalysts in selective transformations, which shows that GR-based composite photocatalysts can be applied as a promising class of material in organic synthesis. The relative reports in this respect demonstrate that coupling GR or GO with the photocatalysts (semiconductors and organocatalysts) is able to change the physiochemical properties (light absorption property, structure and morphology etc.), improve the reaction selectivity and enhance the photocatalytic performance of the as-obtained GR-based composite photocatalysts. Notably, in addition to the widely recognized role of GR/GO as a photoelectron reservoir, the novel functions of GR as an organic dye-like macromolecular photosensitizer, GO as a co-catalyst and GO itself as a photocatalyst for selective organic synthesis, are also demonstrated, which provide new concepts and protocols to the design and preparation of novel GR/GO-based composite photocatalysts for targeted applications in solar energy storage and conversion. The progress made on GR/GO-based composites towards selective organic transformations highlights that there is a wide scope of opportunities in this relatively nascent field. Further attention should be devoted to widening the application of GR-based composite photocatalysts for selective organic transformations and further advancing the possibility of this promising type of composites as the next-generation materials to the practical applications in solar energy storage and conversion.

Acknowledgment

The support by the National Natural Science Foundation of China (NSFC) (Grant No. U1463204, 20903023, 21173045), the Award Program for Minjiang Scholar Professorship and the Natural Science Foundation (NSF) of Fujian Province for Distinguished Young Investigator Grant No. (2012J06003) is gratefully acknowledged.

References

1. A. Fujishima and K. Honda. Electrochemical Photolysis of Water at a Semiconductor Electrode. *Nature* **238**, 1972, 37–38.
2. N. Zhang, R. Ciriminna, M. Pagliaro and Y.-J. Xu. Nanochemistry-Derived Bi_2WO_6 Nanostructures: Towards Production of Sustainable Chemicals and Fuels Induced by Visible Light. *Chem. Soc. Rev.* **43**, 2014, 5276–5287.
3. M. R. Hoffmann, S. T. Martin, W. Choi and D. W. Bahnemann. Environmental Applications of Semiconductor Photocatalysis. *Chem. Rev.* **95**, 1995, 69–96.
4. X. Chen, S. Shen, L. Guo and S. S. Mao. Semiconductor-Based Photocatalytic Hydrogen Generation. *Chem. Rev.* **110**, 2010, 6503–6570.
5. X. Chen, C. Li, M. Gratzel, R. Kostecki and S. S. Mao. Nanomaterials for Renewable Energy Production and Storage. *Chem. Soc. Rev.* **41**, 2012, 7909–7937.
6. Y.-J. and Xu, Y. Zhuang and X. Fu. New Insight for Enhanced Photocatalytic Activity of TiO_2 by Doping Carbon Nanotubes: A Case Study on Degradation of Benzene and Methyl Orange. *J. Phys. Chem. C* **114**, 2010, 2669–2676.
7. N. Zhang, S. Liu and Y.-J. Xu. Recent Progress on Metal Core@Semiconductor Shell Nanocomposites as a Promising Type of Photocatalyst. *Nanoscale* **4**, 2012, 2227–2238.
8. P. Wang, B. Huang, X. Qin, X. Zhang, Y. Dai, J. Wei and M.-H. Whangbo. Ag@AgCl: A Highly Efficient and Stable Photocatalyst Active under Visible Light. *Angew. Chem. Int. Edit.* **47**, 2008, 7931–7933.
9. Z. Yi, J. Ye, N. Kikugawa, T. Kako, S. Ouyang, H. Stuart-Williams, H. Yang, J. Cao, W. Luo, Z. Li, Y. Liu and R. L. Withers. An Orthophosphate Semiconductor with Photooxidation Properties under Visible-Light Irradiation. *Nat. Mater.* **9**, 2010, 559–564.

10. J. C. Colmenares and R. Luque. Heterogeneous Photocatalytic Nanomaterials: Prospects and Challenges in Selective Transformations of Biomass-Derived Compounds. *Chem. Soc. Rev.* **43**, 2014, 765–778.
11. Q. Xiang, J. Yu and M. Jaroniec. Graphene-Based Semiconductor Photocatalysts. *Chem. Soc. Rev.* **41**, 2012, 782–796.
12. G. Palmisano, V. Augugliaro, M. Pagliaro and L. Palmisano. Photocatalysis: A Promising Route for 21st Century Organic Chemistry. *Chem. Commun.* 2007, 3425–3437.
13. J. Liqiang, Q. Yichun, W. Baiqi, L. Shudan, J. Baojiang, Y. Libin, F. Wei, F. Honggang and S. Jiazhong. Review of Photoluminescence Performance of Nano-sized Semiconductor Materials and Its Relationships with Photocatalytic Activity. *Sol. Energ. Mater. Sol. Cells* **90**, 2006, 1773–1787.
14. K. Woan, G. Pyrgiotakis and W. Sigmund. Photocatalytic Carbon-Nanotube–TiO_2 Composites. *Adv. Mater.* **21**, 2009, 2233–2239.
15. J. Yu, T. Ma, G. Liu and B. Cheng. Enhanced Photocatalytic Activity of Bimodal Mesoporous Titania Powders by C_{60} Modification. *Dalton Trans.* **40**, 2011, 6635–6644.
16. Y. Hu, X. Gao, L. Yu, Y. Wang, J. Ning, S. Xu and X. W. Lou. Carbon-Coated CdS Petalous Nanostructures with Enhanced Photostability and Photocatalytic Activity. *Angew. Chem. Int. Edit.* **52**, 2013, 5636–5639.
17. Y. K. Kim and H. Park. Light-Harvesting Multi-Walled Carbon Nanotubes and CdS Hybrids: Application to Photocatalytic Hydrogen Production from Water. *Energ. Environ. Sci.* **4**, 2011, 685–694.
18. W. Chen, Z. Fan, B. Zhang, G. Ma, K. Takanabe, X. Zhang and Z. Lai. Enhanced Visible-Light Activity of Titania via Confinement inside Carbon Nanotubes. *J. Am. Chem. Soc.* **133**, 2011, 14896–14899.
19. R. Leary and A. Westwood. Carbonaceous Nanomaterials for the Enhancement of TiO_2 Photocatalysis. *Carbon* **49**, 2011, 741–772.
20. K. S. Novoselov, A. K. Geim, S. V. Morozov, D. Jiang, Y. Zhang, S. V. Dubonos, I. V. Grigorieva and A. A. Firsov. Electric Field Effect in Atomically Thin Carbon Films. *Science* **306**, 2004, 666–669.
21. X. Huang, X. Qi, F. Boey and H. Zhang. Graphene-Based Composites. *Chem. Soc. Rev.* **41**, 2012, 666–686.
22. D. Chen, H. Feng and J. Li. Graphene Oxide: Preparation, Functionalization, and Electrochemical Applications. *Chem. Rev.* **112**, 2012, 6027–6053.
23. X. An and J. C. Yu. Graphene-Based Photocatalytic Composites. *RSC Adv.* **1**, 2011, 1426–1434.
24. P. V. Kamat, Graphene-Based Nanoassemblies for Energy Conversion. *J. Phys. Chem. Lett.* **2**, 2011, 242–251.

25. M. Pumera. Graphene-Based Nanomaterials for Energy Storage. *Energ. Environ. Sci.* **4**, 2011, 668–674.
26. S. Stankovich, D. A. Dikin, G. H. B. Dommett, K. M. Kohlhaas, E. J. Zimney, E. A. Stach, R. D. Piner, S. T. Nguyen and R. S. Ruoff. Graphene-Based Composite Materials. *Nature* **442**, 2006, 282–286.
27. M. D. Stoller, S. Park, Y. Zhu, J. An and R. S. Ruoff. Graphene-Based Ultracapacitors. *Nano Lett.* **8**, 2008, 3498–3502.
28. N. Zhang, Y. Zhang and Y.-J. Xu. Recent Progress on Graphene-Based Photocatalysts: Current Status and Future Perspectives. *Nanoscale* **4**, 2012, 5792–5813.
29. L. Han, P. Wang and S. Dong. Progress in Graphene-Based Photoactive Nanocomposites as a Promising Class of Photocatalyst. *Nanoscale* **4**, 2012, 5814–5825.
30. Y. Zhang, Z.-R. Tang, X. Fu and Y.-J. Xu. TiO_2–Graphene Nanocomposites for Gas-Phase Photocatalytic Degradation of Volatile Aromatic Pollutant: Is TiO_2–Graphene Truly Different from Other TiO_2–Carbon Composite Materials? *ACS Nano* **4**, 2010, 7303–7314.
31. W. Tu, Y. Zhou and Z. Zou. Versatile Graphene-Promoting Photocatalytic Performance of Semiconductors: Basic Principles, Synthesis, Solar Energy Conversion, and Environmental Applications. *Adv. Funct. Mater.* **23**, 2013, 4996–5008.
32. M.-Q. Yang and Y.-J. Xu. Selective Photoredox Using Graphene-Based Composite Photocatalyst. *Phys. Chem. Chem. Phys.* **15**, 2013, 19102–19118.
33. Y. Shiraishi and T. Hirai. Selective Organic Transformations on Titanium Oxide-Based Photocatalysts. *J. Photochem. Photobiol. C* **9**, 2008, 157–170.
34. M.-Q. Yang, N. Zhang and Y.-J. Xu. Synthesis of Fullerene–, Carbon Nanotube–, and Graphene–TiO_2 Nanocomposite Photocatalysts for Selective Oxidation: A Comparative Study. *ACS Appl. Mater. Interfaces* **5**, 2013, 1156–1164.
35. N. Zhang, Y. Zhang, M.-Q. Yang, Z.-R. Tang and Y.-J. Xu. A Critical and Benchmark Comparison on Graphene-, Carbon Nanotube-, and Fullerene-Semiconductor Nanocomposites as Visible Light Photocatalysts for Selective Oxidation. *J. Catal.* **299**, 2013, 210–221.
36. A. Tanaka, K. Hashimoto and H. Kominami. Preparation of Au/CeO_2 Exhibiting Strong Surface Plasmon Resonance Effective for Selective or Chemoselective Oxidation of Alcohols to Aldehydes or Ketones in Aqueous Suspensions under Irradiation by Green Light. *J. Am. Chem. Soc.* **134**, 2012, 14526–14533.

37. M. Zhang, C. Chen, W. Ma and J. Zhao. Visible-Light-Induced Aerobic Oxidation of Alcohols in a Coupled Photocatalytic System of Dye-Sensitized TiO_2 and TEMPO. *Angew. Chem. Int. Edit.* **47**, 2008, 9730–9733.
38. N. Zhang, X. Fu and Y.-J. Xu. A Facile and Green Approach to Synthesize Pt@CeO_2 Nanocomposite with Tunable Core-Shell and Yolk-Shell Structure and its Application as a Visible Light Photocatalyst. *J. Mater. Chem.* **21**, 2011, 8152–8158.
39. N. Zhang and Y.-J. Xu. Aggregation- and Leaching-Resistant, Reusable, and Multifunctional Pd@CeO_2 as a Robust Nanocatalyst Achieved by a Hollow Core–Shell Strategy. *Chem. Mater.* **25**, 2013, 1979–1988.
40. Y. Zhang, N. Zhang, Z.-R. Tang and Y.-J. Xu. Transforming CdS into an Efficient Visible Light Photocatalyst for Selective Oxidation of Saturated Primary C-H Bonds Under Ambient Conditions. *Chem. Sci.* **3**, 2012, 2812–2822.
41. Y. Zhang, N. Zhang, Z.-R. Tang and Y.-J. Xu. Identification of Bi_2WO_6 as a Highly Selective Visible-Light Photocatalyst Toward Oxidation of Glycerol to Dihydroxyacetone in Water. *Chem. Sci.* **4**, 2013, 1820–1824.
42. Y. Zhang, N. Zhang, Z.-R. Tang and Y.-J. Xu. A Unique Silk Mat-Like Structured Pd/CeO_2 as an Efficient Visible Light Photocatalyst for Green Organic Transformation in Water. *ACS Sustainable Chem. Eng.* **1**, 2013, 1258–1266.
43. F.-X. Xiao, J. Miao, H.-Y. Wang and B. Liu. Self-Assembly of Hierarchically Ordered CdS Quantum Dots–TiO_2 Nanotube Array Heterostructures as Efficient Visible Light Photocatalysts for Photoredox Applications. *J. Mate. Chem. A* **1**, 2013, 12229–12238.
44. F.-X. Xiao, J. Miao, H.-Y. Wang, H. Yang, J. Chen and B. Liu. Electrochemical Construction of Hierarchically Ordered CdSe-Sensitized TiO_2 Nanotube Arrays: Towards Versatile Photoelectrochemical Water Splitting and Photoredox Applications. *Nanoscale* **6**, 2014, 6727–6737.
45. N. Zhang, S. Liu, X. Fu and Y.-J. Xu. A Simple Strategy for Fabrication of "Plum-Pudding" Type Pd@CeO_2 Semiconductor Nanocomposite as a Visible-Light-Driven Photocatalyst for Selective Oxidation. *J. Phys. Chem. C* **115**, 2011, 22901–22909.
46. N. Zhang, S. Liu, X. Fu and Y.-J. Xu. Fabrication of Coenocytic Pd@CdS Nanocomposite as a Visible Light Photocatalyst for Selective Transformation Under Mild Conditions. *J. Mater. Chem.* **22**, 2012, 5042–5052.
47. P. Roy, A. P. Periasamy, C.-T. Liang and H.-T. Chang. Synthesis of Graphene-ZnO-Au Nanocomposites for Efficient Photocatalytic Reduction of Nitrobenzene. *Environ. Sci. Technol.* **47**, 2013, 6688–6695.

48. D. Ravelli, D. Dondi, M. Fagnoni and A. Albini. Photocatalysis. A Multi-Faceted Concept for Green Chemistry. *Chem. Soc. Rev.* **38**, 2009, 1999–2011.
49. A. Maldotti, A. Molinari and R. Amadelli, Photocatalysis with Organized Systems for the Oxofunctionalization of Hydrocarbons by O_2. *Chem. Rev.* **102**, 2002, 3811–3836.
50. G. Palmisano, E. Garcia-Lopez, G. Marci, V. Loddo, S. Yurdakal, V. Augugliaro and L. Palmisano. Advances in Selective Conversions by Heterogeneous Photocatalysis. *Chem. Commun.* **46**, 2010, 7074–7089.
51. N. Sahiner, H. Ozay, O. Ozay and N. A. Aktas. Soft Hydrogel Reactor for Cobalt Nanoparticle Preparation and Use in the Reduction of Nitrophenols. *Appl. Catal. B* **101**, 2010, 137–143.
52. T. Vincent and E. Guibal. Chitosan-Supported Palladium Catalyst. 3. Influence of Experimental Parameters on Nitrophenol Degradation. *Langmuir* **19**, 2003, 8475–8483.
53. J. Du and Z. Xia. Measurement of the Catalytic Activity of Gold Nanoparticles Synthesized by a Microwave-Assisted Heating Method Through Time-Dependent UV Spectra. *Anal. Methods* **5**, 2013, 1991–1995.
54. S. Panigrahi, S. Basu, S. Praharaj, S. Pande, S. Jana, A. Pal, S. K. Ghosh and T. Pal. Synthesis and Size-Selective Catalysis by Supported Gold Nanoparticles: Study on Heterogeneous and Homogeneous Catalytic Process. *J. Phys. Chem. C* **111**, 2007, 4596–4605.
55. M.-Q. Yang, B. Weng and Y.-J. Xu. Synthesis of In_2S_3-CNT nanocomposites for selective reduction under visible light. *J. Mate. Chem. A* **2**, 2014, 1710–1720.
56. M.-Q. Yang, X. Pan, N. Zhang and Y.-J. Xu. A Facile One-Step Way to Anchor Noble Metal (Au, Ag, Pd) Nanoparticles on the Reduced Graphene Oxide Mat with Superb Catalytic Activity for Selective Reduction of Nitroaromatic Compounds. *Cryst. Eng. Comm.* **15**, 2013, 6819–6828.
57. S. U. Sonavane, M. B. Gawande, S. S. Deshpande, A. Venkataraman and R. V. Jayaram. Chemoselective Transfer Hydrogenation Reactions over Nanosized γ-Fe_2O_3 Catalyst Prepared by Novel Combustion Route. *Catal. Commun.* **8**, 2007, 1803–1806.
58. N. Zhang, Y. Zhang, X. Pan, X. Fu, S. Liu and Y.-J. Xu. Assembly of CdS Nanoparticles on the Two-Dimensional Graphene Scaffold as Visible-Light-Driven Photocatalyst for Selective Organic Transformation under Ambient Conditions. *J. Phys. Chem. C* **115**, 2011, 23501–23511.
59. Z. Chen, S. Liu, M.-Q. Yang and Y.-J. Xu. Synthesis of Uniform CdS Nanospheres/Graphene Hybrid Nanocomposites and Their Application as

Visible Light Photocatalyst for Selective Reduction of Nitro Organics in Water. *ACS Appl. Mater. Interfaces* **5**, 2013, 4309–4319.
60. S. Liu, M.-Q. Yang and Y.-J. Xu. Surface Charge Promotes the Synthesis of Large, Flat Structured Graphene-(CdS Nanowire)–TiO_2 Nanocomposites as Versatile Visible Light Photocatalysts. *J. Mater. Chem. A* **2**, 2014, 430–440.
61. F.-X. Xiao, J. Miao and B. Liu. Layer-by-Layer Self-Assembly of CdS Quantum Dots (CdS QDs)/Graphene Nanosheets Hybrid Films for Photoelectrochemical and Photocatalytic Applications. *J. Am. Chem. Soc.* **136**, 2014, 1559–1569.
62. M.-Q. Yang, B. Weng and Y.-J. Xu. Improving the Visible Light Photoactivity of In_2S_3-Graphene Nanocomposite via a Simple Surface Charge Modification Approach. *Langmuir* **29**, 2013, 10549–10558.
63. W. Tu, Y. Zhou, Q. Liu, S. Yan, S. Bao, X. Wang, M. Xiao and Z. Zou. An In Situ Simultaneous Reduction-Hydrolysis Technique for Fabrication of TiO_2–Graphene 2D Sandwich-Like Hybrid Nanosheets: Graphene-Promoted Selectivity of Photocatalytic-Driven Hydrogenation and Coupling of CO_2 into Methane and Ethane. *Adv. Funct. Mater.* **23**, 2013, 1743–1749.
64. S. N. Habisreutinger, L. Schmidt-Mende and J. K. Stolarczyk. Photocatalytic reduction of CO_2 on TiO_2 and other semiconductors. *Angew. Chem. Int. Edit.* **52**, 2013, 7372–7408.
65. W. Tu, Y. Zhou, Q. Liu, Z. Tian, J. Gao, X. Chen, H. Zhang, J. Liu and Z. Zou. Robust Hollow Spheres Consisting of Alternating Titania Nanosheets and Graphene Nanosheets with High Photocatalytic Activity for CO_2 Conversion into Renewable Fuels. *Adv. Funct. Mater.* **22**, 2012, 1215–1221.
66. Y. T. Liang, B. K. Vijayan, K. A. Gray and M. C. Hersam. Minimizing Graphene Defects Enhances Titania Nanocomposite-Based Photocatalytic Reduction of CO_2 for Improved Solar Fuel Production. *Nano Lett.* **11**, 2011, 2865–2870.
67. Y. T. Liang, B. K. Vijayan, O. Lyandres, K. A. Gray and M. C. Hersam. Effect of Dimensionality on the Photocatalytic Behavior of Carbon–Titania Nanosheet Composites: Charge Transfer at Nanomaterial Interfaces. *J. Phys. Chem. Lett.* **3**, 2012, 1760–1765.
68. H.-C. Hsu, I. Shown, H.-Y. Wei, Y.-C. Chang, H.-Y. Du, Y.-G. Lin, C.-A. Tseng, C.-H. Wang, L.-C. Chen, Y.-C. Lin and K.-H. Chen. Graphene Oxide as a Promising Photocatalyst for CO_2 to Methanol Conversion. *Nanoscale* **5**, 2013, 262–268.

69. X. Pan, N. Zhang, X. Fu and Y.-J. Xu. Selective Oxidation of Benzyl Alcohol over TiO_2 Nanosheets with Exposed {0 0 1} Facets: Catalyst Deactivation and Regeneration. *Appl. Catal. A* **453**, 2013, 181–187.
70. S. Higashimoto, N. Suetsugu, M. Azuma, H. Ohue and Y. Sakata. Efficient and Selective Oxidation of Benzylic Alcohol by O_2 into Corresponding Aldehydes on a TiO_2 Photocatalyst Under Visible Light Irradiation: Effect of Phenyl-Ring Substitution on the Photocatalytic activity. *J. Catal.* **274**, 2010, 76–83.
71. U. R. Pillai and E. Sahle–Demessie. Selective Oxidation of Alcohols in Gas Phase Using Light-Activated Titanium Dioxide. *J. Catal.* **211**, 2002, 434–444.
72. D. I. Enache, J. K. Edwards, P. Landon, B. Solsona-Espriu, A. F. Carley, A. A. Herzing, M. Watanabe, C. J. Kiely, D. W. Knight and G. J. Hutchings. Solvent-Free Oxidation of Primary Alcohols to Aldehydes Using Au-Pd/TiO_2 Catal. *Sci.* **311**, 2006, 362–365.
73. R. A. Sheldon, I. W. C. E. Arends, G.-J. ten Brink and A. Dijksman. Green, Catalytic Oxidations of Alcohols. *Acc. Chem. Res.* **35**, 2002, 774–781.
74. D. P. Das, R. K. Barik, J. Das, P. Mohapatra and K. M. Parida. Visible Light Induced Photo-Hydroxylation of Phenol to Catechol Over RGO-Ag_3VO_4 Nanocomposites Without the Use of H_2O_2. *RSC Adv.* **2**, 2012, 7377–7379.
75. N. Zhang, M.-Q. Yang, Z.-R. Tang and Y.-J. Xu. CdS–Graphene Nanocomposites as Visible Light Photocatalyst for Redox Reactions in Water: A Green Route for Selective Transformation and Environmental Remediation. *J. Catal.* **303**, 2013, 60–69.
76. N. Zhang, Y. Zhang, X. Pan, M.-Q. Yang and Y.-J. Xu. Constructing Ternary CdS–Graphene–TiO_2 Hybrids on the Flatland of Graphene Oxide with Enhanced Visible-Light Photoactivity for Selective Transformation. *J. Phys. Chem. C* **116**, 2012, 18023–18031.
77. Y. Zhang, Z.-R. Tang, X. Fu and Y.-J. Xu. Engineering the Unique 2D Mat of Graphene to Achieve Graphene–TiO_2 Nanocomposite for Photocatalytic Selective Transformation: What Advantage does Graphene Have over Its Forebear Carbon Nanotube? *ACS Nano* **5**, 2011, 7426–7435.
78. Y. Zhang, N. Zhang, Z.-R. Tang and Y.-J. Xu. Improving the Photocatalytic Performance of Graphene–TiO_2 Nanocomposites via a Combined Strategy of Decreasing Defects of Graphene and Increasing Interfacial Contact. *Phys. Chem. Chem. Phys.* **14**, 2012, 9167–9175.
79. Y. Zhang, N. Zhang, Z.-R. Tang and Y.-J. Xu. Graphene Transforms Wide Band Gap ZnS to a Visible Light Photocatalyst. The New Role of Graphene as a Macromolecular Photosensitizer. *ACS Nano* **6**, 2012, 9777–9789.

80. N. Zhang, M.-Q. Yang, Z.-R. Tang and Y.-J. Xu. Toward Improving the Graphene-Semiconductor Composite Photoactivity via the Addition of Metal Ions as Generic Interfacial Mediator. *ACS Nano* **8**, 2014, 623–633.
81. L. Yuan, Q. Yu, Y. Zhang and Y.-J. Xu. Graphene-TiO_2 Nanocomposite Photocatalysts for Selective Organic Synthesis in Water Under Simulated Solar Light Irradiation. *RSC Adv.* **4**, 2014, 15264–15270.
82. M.-Q. Yang, Y. Zhang, N. Zhang, Z.-R. Tang and Y.-J. Xu. Visible-Light-Driven Oxidation of Primary C-H Bonds over CdS with Dual Co-catalysts Graphene and TiO_2. *Sci. Rep.* **3**, 2013, 3314–3320.
83. Y. Pan, S. Wang, C. W. Kee, E. Dubuisson, Y. Yang, K. P. Loh and C.-H. Tan. Graphene Oxide and Rose Bengal: Oxidative C-H Functionalisation of Tertiary Amines Using Visible Light. *Green Chem.* **13**, 2011, 3341–3344.
84. S. Wang, C. T. Nai, X.-F. Jiang, Y. Pan, C.-H. Tan, M. Nesladek, Q.-H. Xu and K. P. Loh. Graphene Oxide–Polythiophene Hybrid with Broad-Band Absorption and Photocatalytic Properties. *J. Phys. Chem. Lett.* **3**, 2012, 2332–2336.

Chapter 5

Plasmonic Photocatalysts

Congjun Wang[*,†,‡] and Christopher Matranga[*,§]

*National Energy Technology Laboratory
626 Cochrans Mill Road
Pittsburgh, PA 15236, USA
†AECOM
626 Cochrans Mill Road
Pittsburgh, PA 15236, USA
‡congjun.wang@netl.doe.gov
§christopher.matranga@netl.doe.gov

Plasmonic materials have attracted a tremendous amount of attention because of their fascinating size-, shape- and material-dependent optical properties. These materials have been proposed for use in applications including sensing and imaging, nanooptics, photothermal therapy and catalysis. Recently, there has also been increased interest in utilizing plasmonic materials for energy applications because plasmonic nanostructures have been demonstrated to convert light to electrical, thermal or chemical energy. This chapter focuses on the use of plasmonic materials for photocatalysis applications. The attractiveness of plasmonic photocatalysts arises from the potential of achieving efficiencies and selectivities that are not possible with traditional, thermally-driven, catalytic reactions. The advances in utilizing different plasmonic materials for enhanced photocatalysis and understanding their reaction mechansims are discussed in detail.

1. Introduction

Even though we typically associate plasmonic materials and the control of their optical properties with the explosion of nanotechnology research over the past 30 years, artisans were, unknowingly, utilizing plasmonic nanoparticles for centuries to create vibrantly colored stained glass for use in churches, among other applications.[1] Modern scientific studies of plasmonic materials have led to an in depth understanding of how their optical properties correlate with particle composition, size, shape and structure.[2] The understanding of the physics of these materials combined with the substantial progress in synthesis has enabled the rational design and preparation of a variety of plasmonic material systems with tunable optical properties from the ultraviolet (UV) region well into the near infrared (NIR).[3] The result of the development of these structure-property relationships is exemplified by the technological advancement including sensing at the single molecule level,[4] plasmon-enhanced photovoltaics,[5] plasmonic metamaterials with intriguing optical properties,[6] plasmonic nanophotonics,[7] plasmonic heat generation,[8] photothermal therapy,[9] and the plasmoelectric effect.[10] For photocatalysis, there is a practical need to harness the low energy visible and NIR photons reaching the Earth's surface to drive useful catalytic reactions.[11] As such, an emerging field of materials chemistry and design is developing plasmonic nanostructures specifically for use with either naturally occurring (diffuse)[12] or concentrated (focused) solar light.[13] Although the interest in plasmon-enhanced photocatalysis has grown substantially over the past few years,[14] the notion that plasmonic excitation can enhance photochemistry can be traced further back to theoretical[15] and experimental[16] demonstrations over 30 years ago. In addition, we note that although the term photocatalysis has been widely accepted to describe the various photo-driven chemical conversion processes utilizing plasmonic materials, some of the processes are not strictly photocatalytic by the term's original definition, as illustrated by the discussion of the various mechanisms in this Chapter.

Plasmonic nanostructures are attractive for photocatalysis, particularly for visible-light driven catalysis, largely because of the following properties. First, the localized surface plasmon resonance (LSPR) in metal nanoparticles has very large molar extinction coefficients that are several orders of magnitude greater than those of strongly absorbing organic dye molecules;[17] in addition, the photostability of plasmonic particles is better than many dye systems. Second, the optical properties of these materials can be easily manipulated, in a quantitatively predictable manner, by controlling the composition, size and/or shape.[18] Third, under resonance excitation, the plasmonic nanostructures spatially concentrate the electromagnetic field in their vicinity, generate hot charge carriers, and can also produce heat, all of which can be utilized for enhanced catalysis.[14e] In comparison with conventional heterogeneous catalysts, plasmonic materials offer fascinating possibilities that can enable enhanced performance and potentially controllable reaction pathways based on the energetics of reaction steps.

There has been an exponential rise in the interest for plasmonic photocatalysts since the first demonstration of a complete catalytic cycle using plasmonic nanostructures on an inert support under low intensity visible-light excitation in 2008.[19] The use of plasmonic materials for photocatalysis applications has been demonstrated for a variety of different chemical reactions, such as photo-oxidation of organics,[12c,19,20] Suzuki coupling,[21] water splitting,[22] the visible-light driven photoreduction of CO_2,[13c,23] and plasmon-enhanced chemical vapor deposition.[13b] The reaction mechanisms have also been investigated experimentally and computationally which have shed more light on the fundamental physics and chemistry leading to these processes.[24] As the physical mechanisms leading to chemical reactions become better understood, our ability to design more efficient plasmonic photocatalysts is improved dramatically.

This Chapter will first describe the plasmon excitation and decay processes in plasmonic nanostructures which form the foundation for using these plasmonic materials for photocatalysis applications. We will then review the progress in using plasmonic nanostructures for a variety of photocatalytic reactions, both directly occurring on the surface of the plasmonic nanostructures and on other catalysts, such as semiconductors, in direct contact with the plasmonic materials under excitation. These discussions will be organized based on the three main plasmon excitation driven catalytic mechanisms described in the literature: field enhancement, carrier injection, and plasmonic heating. We conclude the Chapter by offering an overview of possible future endeavors in this field.

2. Plasmonic Nanostructures under Resonance Excitation

The strong interaction between light and plasmonic nanostructures occurs when the frequency of incident photons is in resonance with the coherent oscillation of free electrons. This well-known phenomenon (LSPR) not only results in very interesting optical properties, but excitation of the LSPR initiates intriguing electronic processes that are crucial to photocatalysis applications. The LSPR frequency can be tuned in a wide spectral window including the UV region with Al- and Ag-based materials, the visible region using Au and Cu, and even the NIR by using conducting oxides such as Sn-doped In_2O_3.[3a,3f,18] Additional tunability can be gained by adjusting the size and shape of particles, as well as the dielectric medium the particles are supported in.[2b] The tunability of the plasmon frequency makes these materials immensely attractive for utilizing diffuse or focused sun light to drive photocatalytic processes, as their optical properties can be tuned to match the full solar spectrum through composition, size, and shape.

Plasmonic nanostructures also exhibit exceptionally large extinction cross-sections, typically several orders of magnitude larger than those of strongly absorbing dye molecules.[17] In fact, their extinction cross-section at the LSPR resonance can easily exceed the geometrical cross-section of the nanoparticle itself creating an antennae effect.[25] For example, the extinction cross-section of a small Ag nanoparticle in air is ~10 times the cross-sectional area of the particle, indicating that, to first-order, a surface with 10% coverage of the Ag nanoparticles is sufficient to absorb and scatter all of the incident photons.[26] This property indicates that extremely small quantities of plasmonic materials can have an incredibly large impact on the optical properties and performance of composite or heterostructured systems.

The coherent oscillation of electrons in plasmonic nanostructures under LSPR resonance excitation also results in the concentration of the electromagnetic field in the immediate vicinity of the nanostructures (Fig. 1(a)) as demonstrated by finite-difference time-domain (FDTD) simulations.[14d,25] A field enhancement factor of 10^3 is readily observed near metallic nanoparticles, which can further increase to 10^6 at the so-called "hot spots" that are located between two plasmonic nanoparticles with

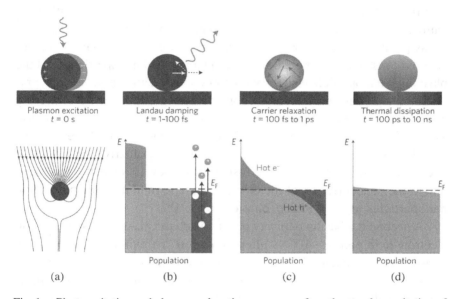

Fig. 1. Photoexcitation and plasmon relaxation processes after a laser pulse excitation of a metallic particle and the associated timescales. (a) The excitation of LSPR concentrates the electromagnetic field. (b)–(d), Schematic representations of the population of the electronic states (grey) following plasmon excitation: hot electrons are represented by the red areas above the Fermi energy E_F and hot hole distributions are represented by the blue area below E_F. (b) In the first 1–100 fs following Landau damping, the athermal distribution of electron–hole pairs decays either through re-emission of photons or through carrier multiplication caused by electron–electron interactions. During this very short time interval τ_{nth}, the hot carrier distribution is highly non-thermal. (c) The hot carriers will redistribute their energy by electron–electron scattering processes on a timescale τ_{el} ranging from 100 fs to 1 ps. (d) Finally, heat is transferred to the surroundings of the metallic structure on a longer timescale τ_{ph} ranging from 100 ps to 10 ns, via thermal conduction. (Reprinted with permission from Ref. 14d. Copyright 2015, Macmillan Publishers Ltd.)

very small separation (~1 nm).[27] To date, the extremely high optical fields occurring from "hot spots" or particle clustering have not been effectively utilized as a rational photocatalysis design tool.[28] For instance, it should be possible to use templated arrays of nanoparticles to create and control the spatial location of these hot spots to further enhance the efficiency of useful photochemistry.[12b,29]

Direct excitation of the LSPR in a nanoparticle deposits energy into the system, which can be dissipated through a variety of decay pathways occurring over femto to nanosecond timescales (Fig. 1).[14d] All of these pathways can directly or indirectly initiate chemical reactions. The first decay pathway to consider is the instantaneous elastic re-radiation of resonant photons, which is more dominant in large metal nanoparticles.[17] If the plasmonic nanoparticle is in proximity or contact with a molecule or semiconductor possessing a resonance at this energy, this elastic scattering can lead to direct photoexcitation of the molecule or semiconductor. The excited charge carriers created by photoexcitation in these systems can then be harvested to initiate a variety of reductive or oxidative chemical events. The photochemistry efficiency is strongly enhanced due to the concentration of the electromagnetic field near the plasmonic nanoparticles under LSPR excitation.[30]

A second dephasing pathway occurs when the plasmon decays into an electron–hole pair ~1–100 fs after excitation. This process is known as non-radiative Landau damping and is more characteristic of smaller nanoparticles.[14c,14d] The electron and hole created are not correlated, as they typically occur in semiconductors, and direct recombination of charge carriers is a rare event. Landau damping results in a population of excited electrons just above the metal Fermi level, E_f, and excited holes below E_f (Fig. 1).[14d] This population does not follow a Fermi–Dirac distribution and is often referred to as "athermal." This distribution of hot electrons can be harvested resulting in the injection of an electron to a molecular state or the conduction band of a semiconductor within the energy range of $E_f + h\nu_{LPSR}$. Likewise, the hot holes create acceptor states for charge transfer of electrons from a molecule to the plasmonic nanoparticle, which can result in oxidation of the molecular species.

The distribution of hot carriers produced by Landau damping eventually thermalizes over a few 100 fs to the well-known Fermi–Dirac distribution.

The thermalized electrons and holes created during this cooling are capable of initiating chemical events similar to those driven by hot carriers before thermalization with the primary difference being the energy of the carriers involved, as well as the time-scales this charge transfer occurs over. As electron-electron interactions cool the electron distribution, these electrons eventually become low enough in energy to couple with phonons and generate heat in the nanoparticle approximately 10–100 ps after plasmonic excitation. Subsequent heat transfer to the environment surrounding the nanoparticle occurs on pico to nanosecond time scales (Fig. 1).[14d] The heat created at the nanoparticle surface and surrounding environment is capable of driving chemical reactions and a few authors have already reported the use of plasmonic heating for the oxidation of CO[31] and reduction of CO_2.[13c]

Another plasmon decay channel becomes possible if adsorbates are present on the surface of the plasmonic particles. This ultrafast (~5fs) energy transfer process is called chemical interface damping (CID) and arises from the coupling of plasmon states with empty electronic states in the adsorbates.[14c] This coupling leads to very rapid (~ 5 fs) charge injection directly from the excited plasmonic nanoparticle to the molecule. Unlike the charge injection process described earlier where energetic carriers are first generated in the plasmonic particle followed by injection into adsorbates, CID occurs instantaneously as the excited plasmon dephases. The primary difference between CID and the other charge transfer process described earlier involves the energetics and time-scales associated with the event. As such, CID is an interesting pathway to consider for further manipulating photocatalyic efficiency and pathways.

3. Plasmonic Photocatalysts

Plasmonic materials have demonstrated a wide range of interesting properties for useful photocatalysis applications. Most well-known plasmonic materials have LSPR in the visible and even NIR, a spectral range where effective photocatalysts would be highly interesting but are currently very scarce.[32] As such, plasmonic photocatalysts offer unique possibilities to impart control of catalytic processes that are impossible to manipulate with conventional semiconductor photocatalysts as well as traditional, thermally-driven catalysts. In this section, we discuss the use of these materials for catalysis applications based on the three main reaction mechanisms proposed in the literature: field enhancement, carrier injection and plasmonic heating. We note that for certain reactions, plasmon enhanced photocatalysis utilizes composite materials where a catalytically active material (such as a semiconductor) is attached to a plasmonic material. In these materials, the plasmonic material transfers energy under resonance excitation to the catalytically active material where catalysis reaction takes place. Alternatively, there have been many reports on photocatalysis directly driven by plasmonic materials. In these systems, the plasmonic nanostructures initiate catalysis reactions directly on their surface under illumination without the need for a separate catalytic material. Furthermore, plasmonic nanoparticles such as Au have been reported to enhance photocatalysis under interband excitation.[23,33] Since these processes are not directly related to plasmon excitation, they are not discussed here. Finally, although the rest of this section is discussed based on the reaction mechanisms occuring, it is important to emphasize that in many cases, several different plasmonic enhancement mechanisms work synergestically together to achieve the observed enchanced performance.

3.1 Plasmon enhanced photocatalysis via field enhancement

A variety of photocatalysts, mostly based on semiconductors and molecular catalysts, has been extensively studied for decades. However, the efficiency of these catalysts, particularly under low energy visible light irradiation, remains low and impractical. Therefore, there has been significant interest in utilizing the optical properties of plasmonic materials to enhance the performance of catalytically active materials. One strategy is to utilize the strong field enhancement effect of plasmonic nanostructures under LSPR excitation to improve the photocatalytic activity of a material in close proximity.

The attenae effect of plasmonic nanostructures for concentrating the eletromagnetic field under resonance excitation is well-known. Simulation of the electical field distribution around different plasmonic nanostructures and plasmonic/semiconductor interfaces has been performed by different groups using the well established FDTD method. This is also a straightforward approach for using plasmonic materials to enhance the activity of catalysts that are in the vicinity of the plasmonic nanostructures. Under resonance excitation, the field can easily be enhanced by 10^3 or more. This is equivalent to illuminating the photocatalytic semiconductor, such as TiO_2, near the plasmonic nanostructures with a light intensity that is several orders of magnitude higher than that of the light source. This effective increase in the excitation intensity leads to improved efficiency.

A large amount of work has been reported on the use of the Au/TiO_2 system for enhanced photocatalysis based on the field enhancement effect,[34] a result of the fact that TiO_2 is probably the most extensively studied semiconductor photocatalyst. Using this strategy, Au/TiO_2 based heterostructured plasmonic photocatalysts have been utilized for decomposition of organic compounds,[30] photocatalytic water splitting (Fig. 2(a))[22a] and CO_2 reduction (Fig. 2(b)).[23]

Since this mechanism requires the direct absorption of photons by the catalytically active semiconductor and the fact that the bandgap of TiO_2 is much larger than the plasmon resonance energy of Au, this only works because of the presence of defect states on TiO_2 allowing sub-bandgap

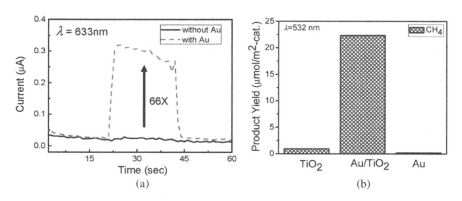

Fig. 2. (a) Plasmon resonant enhancement of photocatalytic water splitting under visible illumination: photocurrent of anodic TiO_2 with and without Au nanoparticles irradiated with $\lambda = 633$ nm light for 22 s. (Reprinted with permission from Ref. 22(a). Copyright 2011, American Chemical Society.) (b) Product yields of the photocatalytic reduction of CO_2 (after 15 h of visible irradiation) on three different catalytic surfaces. (Reprinted with permission from Ref. 23. Copyright 2011, American Chemical Society.)

light absorption. This approach is therefore advantageous because it enables large bandgap TiO_2 to more effectively use low energy visible photons for photocatalysis. The importance of the matching of the LSPR absorption with the absorption of semiconductor photocatalysts is illustrated by the observation that Ag cubes supported on nitrogen-doped TiO_2 (N–TiO_2) exhibit the highest photocatalytic activity enhancement for the decomposition of methylene blue (MB) because of the much larger overlap between the plasmon resonance of Ag cubes and the absorption of N–TiO_2 (Fig. 3).[35]

In addition to TiO_2, a similar strategy has been succesfully demonstrated using other catalysts such as Au/Fe_2O_3 (Fig. 4(a))[12b] and Ag/SiO_2/CdS (Fig. 4(b))[36] for photocatalytic water splitting. It has been shown that for Ag/SiO_2/CdS, the photocatalytic activity is strongly dependent on the distance between the plasmoinc Ag and the semiconductor CdS, which further supports the near-field electromagentic enhancement mechanism.[36]

As noted earlier, in many of these examples, multiple processes simultaneously contribute to the enhanced photocatalytic activity and

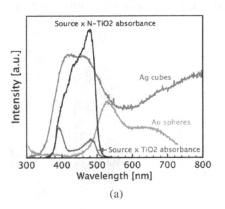
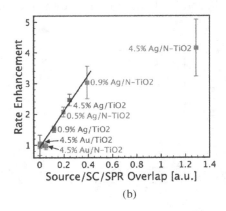

(a)   (b)

Fig. 3. (a) Source/semiconductor absorbance overlap for N–TiO_2 (black curve) and TiO_2 (purple curve) and metal nanoparticle spectra for Ag/N–TiO_2 composite (4.5% Ag) and Au/N–TiO_2 composite (4.5% Au). (b) Enhancement in the MB decomposition rate as a function of the overlap among illumination source, semiconductor absorbance and metal nanoparticle SPR. Blue points shows native TiO_2-based composites, and red points show N–TiO_2-based composites. (Reprinted with permission from Ref. 35. Copyright 2011 American Chemical Society.)

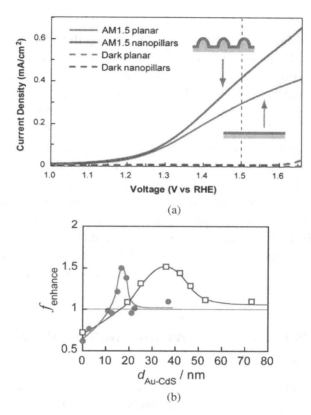

Fig. 4. (a) Enhanced photocurrent from 90 nm Fe_2O_3 (red color in the schematic) coated on an Au nanopillar array (gold color in the schematic). The I–V curves show enhancement of the photocurrent from the patterned Fe_2O_3 electrode as compared to the planar control under AM1.5 simulated solar illumination. (Reprinted with permission from Ref. 12b. Copyright 2012, American Chemical Society.) (b) Experimentally obtained dependence of the enhancement factor, $f_{enhance}$, for H_2 evolution rate on the distance between CdS and Au, d_{CdS-Au}, for CdS@SiO_2-deposited Au@SiO_2 particles with Au cores of 19 nm (solid circles) or 73 nm (open squares). (Reprinted with permission from Ref. 36. Copyright 2011, American Chemical Society.)[36]

the positive role of plasmonic materials is more than just field enhancement. For example, plasmonic metal nanoparticles also serve as an electron reservoir as the metallic particles accept the photoexcited electrons from the conduction band of the semiconductors. This process leads to reduced carrier recombination in the semiconductor catalysts, which is beneficial to photocatalysis, and is evidenced by the observed blue-shift of the LSPR frequency.[37] In addition, the accumulation of electrons in metal nanoparticles undergoes charge equilibration with the photoexcited semiconductor catalysts driving the Fermi level to more negative potentials,[38] which makes the carriers more energetically active for catalysis.

Another positive consequence of local field enhancement in plasmonic material/catalytic semiconductor heterostructure is that the plasmonic particles can very effectively couple light from the far-field to the near-field.[14a] The result of this is that the photogenerated carriers are much closer to the catalyst surface where they more effectively contribute to catalysis. In contrast, when plasmonic nanoparticles are not present, carrier recombination in the bulk of the semiconductor is dominant due to the mismatch of light absorption length and the minority carrier diffusion length.[39]

3.2 LSPR induced carrier injection mediated photocatalysis

As the excited plasmon decays in nanoparticles, particularly in small nanoparticles (for Ag, < 30 nm),[40] energetic carriers are generated via nonradiative Landau damping. In the presence of a catalytically active semiconductor in direct contact with the plasmonic particles, the energetic carriers can inject into the semiconductor (Fig. 5)[14b] where they initiate catalytic reactions.

The first hint of plasmon induced charge separation in noble metal particles in contact with TiO_2 was reported in 1996.[41] Numerous subsequent studies have demonstrated the hot carrier generation in plasmonic nanostructures and injection into semiconductors under LSPR resonance excitation. This phenomenon has also been utilized for photovoltaic and photocatalysis applications for a wide range of materials such as Au or Ag nanoparticles in contact with various TiO_2 structures,[35,42] Au–ZnO,[43] Au–CeO_2,[44] Ag-AgBr,[45] Ag–AgCl,[12c,46] Ag–AgI,[47] etc. Charge transfer from other metals such as Pd and Pt has also been reported.[48]

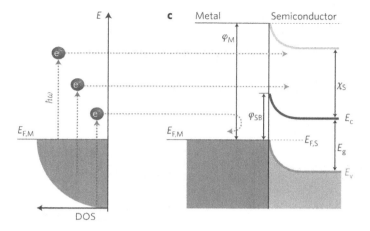

Fig. 5. Electrons from occupied energy levels are excited above the Fermi energy as the plasmon decays in a plasmonic nanoparticle. Hot electrons can be injected into a semiconductor by forming a Schottky barrier with the plasmonic nanostructure. Hot electrons with energies high enough to overcome the Schottky barrier $\varphi_{SB} = \varphi_M - \chi_S$ are injected into the conduction band E_c of the neighbouring semiconductor, where φ_M is the work function of the metal and χ_S is the electron affinity of the semiconductor. (Reprinted with permission from Ref. 14b. Copyright 2014, Macmillan Publishers Ltd.)

The kinetics of the hot carrier generation, injection and recombination processes have been investigated using Au–TiO$_2$ as a model system[49] (Fig. 6(a)). The increased visible and IR absorption of TiO$_2$ after electron injection into its conduction band allows ultrafast spectroscopy techniques to monitor these processes (Fig. 6(b)). The hot electron generation and injection are found to complete within ~50 fs upon LSPR excitation. The electron injection efficiency is dermined to be ~40% for Au–TiO$_2$ under 550 nm excitation. In the presence of electron donors such as I$^-$, the oxidized Au nanoparticles regenerate within ~20 ns. In a different study, the quantum yield of hot electron injection from Au into CdS nanorods is found to be ~2.75%, with interesting potential for further improvement because the electronic properties of quantum confined CdS nanorods can be controlled by their size and shape.[50]

Hot electron injection has been utilized for a variety of plasmon enhanced photocatalysis processes. This is of particular interest because

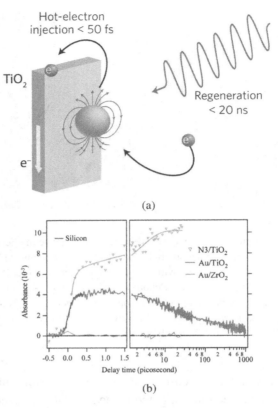

Fig. 6. (a) Schematics of the timescale of the hot electron injection and regeneration processes. (Reprinted with permission from Ref. 14b. Copyright 2014, Macmillan Publishers Ltd.) (b) Ultrafast visible-pump/infrared-probe femtosecond transient absorption spectroscopy shows the kinetics of the hot-electron injection in the TiO_2 conduction band from Au nanoparticles, as well as comparison with ruthenium N3 dye on TiO_2, which is known to have a carrier injection efficiency of almost 100%, and Au on ZrO_2 where no carrier transfer occurs. (Reprinted with permission from Ref. 49b. Copyright 2007, American Chemical Society.)

it provides another means to enable catalytically active semiconductor catalysts with large bandgaps, such as TiO_2 and ZnO, to drive photocatalytic reactions under visible-light illumination. An example is the use of Au–ZnO photoelectrodes for photocatalytic water splitting under simulated solar irradiation (Fig. 7).[43b] The enhancement effect due to hot electron injection from Au into ZnO is investigated by the analysis of the

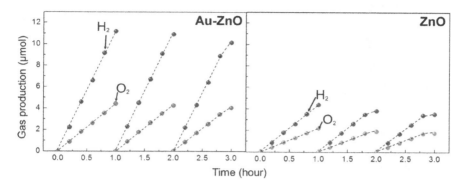

Fig. 7. Time courses of H_2 and O_2 evolution using Au–ZnO and ZnO photoelectrodes under AM 1.5G solar simulator in 0.5 M Na_2SO_4 aqueous solution. (Reprinted with permission from Ref. 43b. Copyright 2012, American Chemical Society.)

wavelength dependent photocurrent in the photoelectrochemical cell, showing a significant contribution of hot electron injection from Au under LSPR excitation.

More recently, it has been reported by two different groups demonstrating the use of plasmon generated hot electrons and holes in the same photocatalytic system for water splitting where all charge carriers are from the resonant plasmon excitation (Fig. 8).[51] In one example (Fig. 8(a)), an array of aligned Au nanorods is capped with crystalline TiO_2 which serves as an electron filter. The TiO_2 layer is further decorated with Pt nanoparticles. The side of each Au nanorod is deposited with a Co oxygen evolution catalyst (Co-OEC).[51a] Upon resonance excitation of the Au nanorods, the hot electrons are injected over the Schottky barrier into the conduction band of TiO_2. These electrons then migrate to the Pt nanoparticles where hydrogen production occurs. In the meantime, the hot holes are transported to Co-OEC to drive oxygen evolution reactions, completing the full photocatalytic water splitting cycle (Fig. 8(b)).

Another example is based on Au nanoparticles loaded on the front side of a single crystal Nb–$SrTiO_3$ (Fig. 8(c)).[51b] Pt co-catalysts are attached to the back side of the same Nb–$SrTiO_3$ crystal for H_2 evolution. Upon photoexcitation of the Au nanoparticles, electrons are injected into the conduction band of Nb–$SrTiO_3$, followed by transfer to the Pt on the

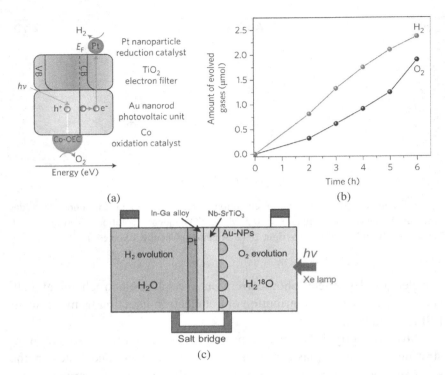

Fig. 8. (a) Energy level diagram superimposed on a schematic of an individual unit of the plasmonic solar water splitter, comprised of the gold nanorod, the TiO_2 cap decorated with platinum nanoparticles, which functions as the hydrogen evolution catalyst, and the Co-OEC material deposited on the lower portion of the gold nanorod, showing the proposed processes occurring in its various parts and in energy space. CB, conduction band; VB, valence band; E_F, Fermi energy. (Reprinted with permission from Ref. 51a. Copyright 2013, Macmillan Publishers Ltd.) (b) Measured O_2 and H_2 photoproducts as a function of time for a device shown in (a) illuminated by 300 mWcm^{-2} of white light (AM 1.5). The hydrogen/oxygen ratio is ~2, trending downward as the experiment progresses, due to contamination by atmospheric oxygen during the extraction of the gaseous products by syringe through a septum. (Reprinted by permission from Ref. 51a. Copyright 2013, Macmillan Publishers Ltd.) (c) A schematic illustration of the water splitting device using the Au-nanoparticle-loaded Nb–SrTiO$_3$ photoelectrode. (Reprinted with permission from Ref. 51b. Copyright 2014, John Wiley & Sons, Inc.)

back side for H_2 generation. The holes trapped near the Au/SrTiO$_3$/water interface oxidize OH$^-$ via multiple-electron transfer to produce oxygen. This device is robust and even active under 750 to 850 nm light irradiation, demonstrating remarkable potential for solar driven catalytic water splitting.

Different studies have also been carried out to utilize the direct carrier injection into adsorbates on the surface of plasmonic nanoparticles to drive catalytic reactions. Early studies into this reaction pathway investigated electron driven chemical reactions on flat, single-crystal, metal surfaces under intense laser excitation.[52] These studies clearly demonstrate reaction pathways that are distinct from thermally driven processes. These model single crystal metal surfaces, however, are not practical for real world catalysis applications. In addition, the observation that electron induced processes occur after the formation of a high effective temperature Fermi–Dirac distribution on extended metal surfaces means that only molecular states close to the Fermi level of the metal can accept these excited electrons.[53] This indicates that, as a practical approach, the selective activation of specific adsorbate resonances is challenging on extented metal surfaces and likely not practical for larger scale catalysis applications.[14d] The lifetimes of the hot electrons, however, can be significantly longer on metal nanoparticles due to increased confinement, more granular density of states and reduced electron-electron interactions.[24b,54] Moreover, decreased electron-phonon coupling in nanometer scale metallic particles results in increased equilibration time with the lattice.[55] Consequently, there has been a strong interest in utilizing metal nanoparticles to catalyze electron driven chemical transformations.

The catalytic conversion of HCHO to CO_2 on Au nanoparticles under low intensity red light illumination of the plasmon resonance clearly demonstrates that direct photocatalysis on plasmonic nanoparticles can be achieved (Fig. 9(a)).[19] Although this result was attributed to a thermal effect when first reported, subsequent experiments indicate that the observed catalytic activity is associated with electronic driven processes, as the photothermal effect is eliminated due to the low light excitation intensity.

A variety of plasmon-enhanced, electron driven, catalytic reactions under low intensity light excitation have since been reported along with some rather detailed studies of the reaction mechanisms involved with these reactions. One interesting study reports the hot electron mediated dissociation of H_2 on Au/TiO_2 under low intensity visible-light excitation

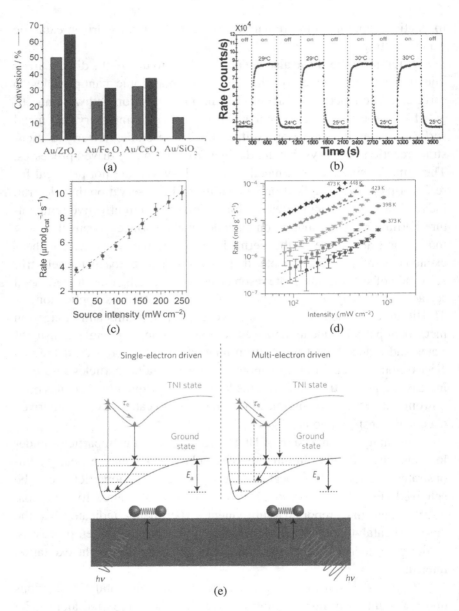

Fig. 9. (a) HCHO conversion activity using Au nanoparticles on different support under blue (blue bars) and red (red bars) light illumination. (Reprinted with permission from Ref. 19. Copyright 2008, John Wiley & Sons, Inc.) (b) Plasmon induced dissociation rate of H_2/D_2 over Au/TiO_2 with laser excitation (on 2.41 W/cm²) and without laser excitation (off).

Fig. 9. (*Continued*) (Reprinted with permission from Ref. 56. Copyright 2013, American Chemical Society.) (c) The rate of ethylene epoxidation as a function of light intensity using Ag nanocubes. (Reprinted with permission from Ref. 57. Copyright 2011, Macmillan Publishers Ltd.) (d) Photocatalytic ethylene epoxidation rate as a function of light intensity at different temperatures. (Reprinted with permission from Ref. 58. Copyright 2012, Macmillan Publishers Ltd.) (e) Left: the molecular mechanism in the linear regime. A single electron excitation deposits vibrational energy into the adsorbate by accelerating the molecule along the TNI potential energy surface (PES) for the lifetime, τ_e. If vibrational energy is not sufficient to overcome the activation barrier, E_a, the adsorbate returns to the thermally equilibrated state. Right: in the super-linear regime the adsorbate is electronically excited multiple times before overcoming the activation barrier. (Reprinted with permission from Ref. 58. Copyright 2012, Macmillan Publishers Ltd.)[58]

at room temperature (Fig. 9(b)).[56] The effect of plasmonic heating was ruled out as a reaction mechanism by careful use of control experiments in the dark. The central role of Au is further confirmed by subsequent studies demonstrating even higher activity when the inert SiO_2 is used to replace the TiO_2 substrate.[55] This is because, unlike TiO_2 which is an electron acceptor competing with the electron injection into H_2 molecule, electron transfer from Au to SiO_2 is no longer possible, leading to enhanced hot electron mediated H_2 dissociation.

A very distinct characteristic of the electronic nature of these reactions is the linear or super-linear reaction rate dependence on light intensity, in contrast to the exponential dependence for thermally driven reactions. This linear dependence has been demonstrated for different reaction systems, and one such example is the ethylene epoxidation on Ag nanocubes deposited on Al_2O_3 support (Fig. 9(c)).[57] At higher light intensities, the linear dependence transitions to a super-linear dependence (Fig. 9(d)).[58] The association of the observed linear and super-linear light intensity dependence with electronic processes is futher supported by isotopic labeling experiments[59] measuring the kinetic isotope effect (KIE).[58] In the linear dependence regime, each reaction is driven by a single photon absorption event and the subsequent interaction between the excited charge carrier with the reactant molecule. At higher light excitation intensity, as the electron injection rate increases, the electronic excitation of an already excited molecule becomes substantial, resulting in the transition to the super-linear regime (Fig. 9(e)).[58]

A very exciting possibility for plasmonic photocatalysts is to selectively deposit energy into specific transient negative ion (TNI) states so that the reaction pathway can be rationally controlled, which is impossible using conventional, thermally driven catalysts. Another interesting opportunity for controlling the reaction selectivity is demonstated by using LSPR to induce the change of the oxidation state of the surface of Cu-based nanoparticles for propylene epoxidation.[13a] The selectivity for propylene oxide (PO) is enhanced by exciting the LSPR of Cu catalysts using visible-light which leads to the reduction of Cu_2O on the surface to form metallic Cu.

To further increase charge separation in plasmonic Au/TiO_2 photocatalysts, mesocrystalline TiO_2 superstructures are used to promote charge migration from the Au/TiO_2 interface to the edges of the mesocrystalline TiO_2 where photocatalysis reaction occurs (Fig. 10(a)).[60] This strategy indeed leads to increased electron lifetime and an enhancement of photocatalytic activity by an order of magnitude for organic pollutant degradation (Fig. 10(b)).[60]

The successful demonstration of hot carrier driven photochemistry using plasmonic photocatalysts has led to increased interest in the theoretical chemistry community to study the properties of hot carriers in plasmonic nanoparticles.[24b,61] A better understanding of these hot carriers could enable interesting opportunities to control the reaction chemistry by, for example, optimizing the generation of hot carriers at specific energies. Despite the significant promise of plasmon enhanced, electron-driven photocatalysis, substantial challenge remains for practical applications. Although the reported reaction rates range from μmol per gram of catalyst per h[60] to mmol per gram of catalyst per h,[13a,51a,58] and the reported apparent quantum efficiency can also be from $3 \times 10^{-4}\%$[60] to 60%[58] depending on specific reactions and catalysts, large scale practical application using carrier injection initiated catalysis requires the reaction rates to be improved by several orders of magnitude, particularly under natrual or concentrated sunlight illumination. In addition, for many applications, hole scavengers need to be added to regenerate oxidized plasmonic nanoparticles which may present additional challenges.[49a,62]

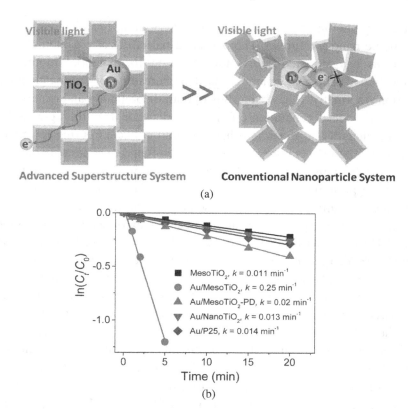

Fig. 10. (a) Schematic of the Au/Meso TiO_2 superstructure compared with the disordered Au/Nano TiO_2 illustrating the enhanced charge separation enabled by the ordered TiO_2 nanocrystal networks in Au/Meso TiO_2. (b) Kinetic linear fitting curves for liquid-phase photocatalytic degradation of the dye MB over different samples under visible-light irradiation (460–700 nm). (Meso TiO_2: pure mesocrystalline TiO_2 superstructure; Au/Meso TiO_2: Au nanoparticles/Meso TiO_2; Au/Meso TiO_2–PD: Photodeposited Au nanoparticles on Meso TiO_2; Au/Nano TiO_2: Au nanoparticles on disordered TiO_2 nanocrystals; Au/P25: Au nanoparticles/Degussa P25 TiO_2. (Reprinted with permission from Ref. 60. Copyright 2014, American Chemical Society.)

3.3 Plasmonic heating for photocatalysis

As described earlier, the relaxation of excited plasmons eventually results in localized heating of the metallic nanostructures.[8b,8c] The efficiency of converting photon energy to heat can be very high, reaching >95%, for plasmonic nanostructures.[3e] The photothermal effect of plasmonic nanostructures has attracted substantial attention for potential biomedical applications.[9b] The heat generation is also interesting for driving chemical reactions thermally. It has been shown, both theoretically and experimentally, that the heat generation efficiency is dependent on the light excitation wavelength and maximum heat generation occurs during resonance excitation. The temperature increase has been found to be linearly dependent on the excitation intensity. The heat generation is also sensitive to the size of the plasmonic nanostructures, the dielectric properties of the surrounding medium, as well as the assembly of plasmonic nanostructures. For example, larger plasmonic nanoparticles are expected to generate more heat.[8b] However, as the size gets larger, light scattering becomes more dominant which reduces the overall light to heat conversion efficiency.

The use of plasmonic heating to thermally drive catalytic reactions is advantageous.[63] First, the energy efficiency can be easily over 10^5 higher compared to conventional thermal reactions where the entire reactor needs to be heated to high temperatures. The light intensity needed to drive these reactions is relatively modest and potentially achievable by concentrated solar illumination. Second, plasmonic heating enables an extremely high degree of spatial control of the nanoscale thermal enviroments which can initiate thermally driven reactions at arbitrarily specified locations at single nanostructure level in a room-temperature reactor. This highly locazlied heating nature further eliminates the need for heat management often required for conventional thermally driven catalytic processes and reactor systems. Third, plasmonic heating also allows temporal control of the heating process at timescales easily in the range of 10–100 μs and even as short as 10 ps to 1 ns. Finally, the plasmonic heating process does not require the generation of charge carriers for photocatalysis, which can potentially improve overall efficiency and erase the photocorrosion problem for various photocatalysts. These intriguing properties therefore open up opportunities that are not possible with traditional thermal catalysts such as the precise control of the reaction process and product selectivity. With the high photon to heat conversion efficiency demonstrated for these materials and the compatibility of these processes with existing thermally driven catalysis industry, large scale deployment of these plasmonic heating enhanced catalysts is promising.

The use of plasmonic heating to drive catalytic reaction was first demonstrated in several plasmon-assisted chemical vapor deposition processes using Au nanoparticles[13b] (Fig. 11(a)) to prepare a variety of materials, such as PbO, TiO_2, Si nanowires[63] (Figure 11(b)) and carbon nanotubes.[64] The temperature increase under plasmon excitation in these experiments is consistent with theoretical prediction (Fig. 11(c)) as well as the experimentally observed plasmonically-initiated growth of Si nanowires on the

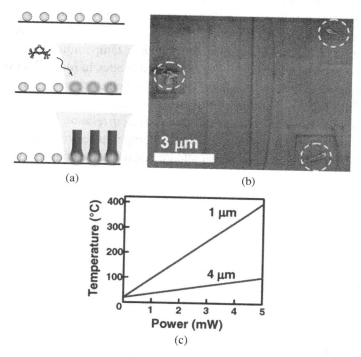

Fig. 11. (a) Schematic of PACVD process. A template of nanoscale gold lines or dots is first laid down on a substrate (top). The substrate is then exposed to a gaseous environment containing the CVD precursor in a carrier gas, and a laser (green) is focused on the surface heating the nanoparticles (middle). Growth is initiated only on the heated nanoparticles (bottom). (Reprinted with permission from Ref. 13b. Copyright 2006, American Chemical Society.) (b) SEM image of three patterned Si nanowires (located inside the white circles) (Reprinted with permission from Ref. 63. Copyright 2007, American Chemical Society.) (c) Calculated temperature in the center of a gold nanoparticle layer illuminated with a 532 nm laser as a function of excitation power. Calculations for two illumination spot radii, r_0, of 1 and 4 µm are shown. (Reprinted with permission from Ref. 63. Copyright 2007, American Chemical Society.)

Au nanoparticles which requires a threshold temperature of ~380°C. Plasmonic heating is also reported to increase the reaction rate between hexacyanoferrate-(III) and thiosulfate in solution[65] and to catalyze steam reforming of ethanol in a microfluidic channel, in which the plasmonic heating not only provides the heat needed for the chemical reaction but also generates water and ethanol vapor locally over the catalysts.[66]

More recently, Au–ZnO heterostructured catalysts have been demonstrated to catalyze the thermal conversion of CO_2 and H_2 to form CH_4 and CO under 532 nm laser excitation (Fig. 12).[13c] In this work, researchers used a heating stage and Raman spectroscopy with a 532 nm laser excitation source to establish a calibration between temperature and the linewidths and resonance frequencies of phonon modes in plain ZnO without plasmonic Au particles attached. The researchers then used the 532 nm laser of the Raman spectrometer to collect intensity-dependent spectra of the plasmonic Au–ZnO system and establish a correlation between excitation intensity and the average ZnO lattice temperature probed by the laser spot. This simple characterization allowed quantitative characterization of

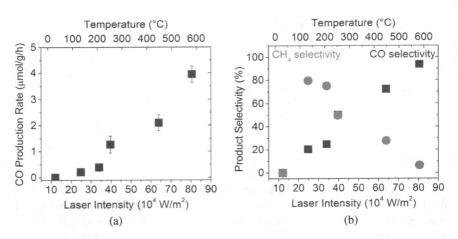

Fig. 12. (a) CO production rate as a function of cw 532 nm laser intensity using Au/ZnO catalysts. (b) Experimental CO (blue squares) and CH_4 (red circles) selectivity as a function of cw 532 nm laser intensity and the corresponding catalyst temperature using Au/ZnO catalysts which is qualitatively consistent with the thermodyanamically simulated selectivity of the thermally driven CO_2 reduction process. (Reproduced with permission from Ref. 13c. Copyright 2013, The Royal Society of Chemistry.)[13c]

how heat flows from the Au nanoparticles to the ZnO substrate during laser excitation and improved the correlation of catalytic properties with reaction thermodynamics. The 532 nm laser source was fully resonant with the LPSR of the Au nanoparticles on the ZnO and local temperature increases on the ZnO substrate approach ~600°C at the highest laser excitation intensity. In this system, the resonantly excited Au nanoparticles convert photon energy to heat, and the subsequent heat transfer to catalytically active ZnO results in the thermal reduction of CO_2. The detected products include CO and CH_4. The importance of the presence of both optically active Au and catalytically active ZnO is confirmed by control experiments indicating that neither Au nor ZnO alone is active for the CO_2 reduction reaction. In addition to the temperature increase measured by Raman spectroscopy of the catalyst under resonance excitation, the thermally driven mechanism is further supported by the following observations. First, the dependence of CO production rate on laser intensity follows an exponential relationship (Fig. 12(a)), which is characteristic of thermally driven process. Second, the laser intensity dependent product distribution is consistent with the thermodynamically predicted temperature dependent product selectivity. Third, the reaction proceeds in a similar manner for experiments carried out under 532 nm laser excitation and those heated thermally in the dark.

As mentioned earlier, in many cases, several plasmon enhancement mechanisms work concertedly to improve the efficiency of photocatalysis reactions.[31,57] One such example is the use of Au–Pd nanostructures to catalyze Suzuki coupling reactions under visible and NIR illumination.[21] The plasmonic heating effect is unambiguously measured in these experiments and the solution temperature increases to ~70°C within 20 min under 1.68 W 809 nm laser illumination with >95% light to heat conversion efficiency (Table 1). Comparison of the Suzuki coupling reaction results from Au–Pd nanostructures under 809 nm laser illumination and those from isothermal heating is summarized in Table 1.[21] As the laser power increases, the reaction yield under NIR illumination significantly exceeds that of the reaction carried out at the same temperature under isothermal heating. This result indicates that both plasmonic heating and plasmonic hot electron generation contribute to the enhanced activity of the Au–Pd nanostructures under laser illumination.[21] Additional analysis

Table 1: Yields of the suzuki coupling reaction between bromobenzene and m-tolylboronic acid in the presence of the large Au–Pd nanostructures under the 809 nm laser illumination and isothermal heating. (Reprinted with permission from Ref 21. Copyright 2013, American Chemical Society.)[21]

	Under laser illumination[a]			Under isothermal heating[b]	
Laser power (W)	End temperature (°C)	Photothermal conversion efficiency (%)	Yield (%)	Heating temperature (°C)	Yield (%)
1.68	62	95.8	99	62	55
1.30	56	96.8	70	56	37
0.89	45	87.9	34	45	33

[a]Environmental temperature, 25°C, reaction time, 20 min. [b]Reaction time, 20 min.

demonstrates that plasmonic heating dominates under low laser power illumination and as laser power increases, the increased plasmon excitation induced hot electron generation becomes much more relevant.[21] Finally, the reaction yield also reaches 99% using Au–Pd nanostructures under natural solar irradiation (Table 2), illustrating the potential of the direct utilization of sunlight for a variety of industrially important catalytic reactions with plasmonic photocatalysts.[21]

These experimental demonstrations provide clear evidence of the many advantages of using plasmonic heating to drive catalytic reactions. Future studies will expand the types of catalytic processes to include other industrially important reactions. Another interesting focus will be to test these processes at a larger scale and further study the use of natural or concentrated solar illumination as the light source.

Table 2. Yields of the Suzuki Coupling Reactions in the Presence of the Mixture of the Medium, Small, and Spherical Au–Pd Nanostructures under solar radiation.[a] (Reprinted with permission from Ref. 21. Copyright 2013, American Chemical Society.)

entry	halide	boronic acid	yield [%]	entry	halide	boronic acid	yield [%]
1	Ph–I	Ph–B(OH)$_2$	97	8	2-OCH$_3$-C$_6$H$_4$–Br	4-CH$_3$-C$_6$H$_4$–B(OH)$_2$	70
2	Ph–Br	Ph–B(OH)$_2$	99	9	2-OCH$_3$-C$_6$H$_4$–Br	2-CH$_3$-C$_6$H$_4$–B(OH)$_2$	99
3	Ph–Br	2-CH$_3$-C$_6$H$_4$–B(OH)$_2$	99	10	2-OCH$_3$-C$_6$H$_4$–Br	3-CH$_3$-C$_6$H$_4$–B(OH)$_2$	99
4	Ph–Br	3-CH$_3$-C$_6$H$_4$–B(OH)$_2$	96	11	2-OCH$_3$-C$_6$H$_4$–Br	4-CH$_3$-C$_6$H$_4$–B(OH)$_2$	99
5	Ph–Br	4-CH$_3$-C$_6$H$_4$–B(OH)$_2$	99	12	4-CH$_3$O-C$_6$H$_4$–Br	2-CH$_3$-C$_6$H$_4$–B(OH)$_2$	99
6	2-OCH$_3$-C$_6$H$_4$–Br	2-CH$_3$-C$_6$H$_4$–B(OH)$_2$	50	13	4-CH$_3$O-C$_6$H$_4$–Br	3-CH$_3$-C$_6$H$_4$–B(OH)$_2$	76
7	2-OCH$_3$-C$_6$H$_4$–Br	3-CH$_3$-C$_6$H$_4$–B(OH)$_2$	67	14	4-CH$_3$O-C$_6$H$_4$–Br	4-CH$_3$-C$_6$H$_4$–B(OH)$_2$	55

[a]Au–Pd nanostructure mixture solution, 1 mL; NaOH, 10 mg; CTAB, 36 mg; H$_2$O, 1 mL; reactant amounts of entries 1–5, 0.08 mmol halide and 0.08 mmol boronic acid; reactant amounts of entries 6–14, 0.04 mmol halide and 0.04 mmol boronic acid; solar power, 565 W·m^{-2}; environment temperature, 30°C; reaction time, 2 h.

4. Conclusion and Outlook

With the advancement of synthesis techniques and the understanding of the physics of plasmonic nanostructures, utilizing these materials' fascinating properties for enhanced photocatalysis has become increasingly interesting and potential breakthrough in accomplishing catalytic processes with controllable reaction pathways is no longer imaginary.

A wide variety of plasmon enhanced photocatalytic processes has been accomplished utilizing the field enhancement, hot carrier injection and plasmonic heating processes. Distinct reaction mechanisms have been investigated and understanding of the physics involved in these processes has enabled improvement of the photocatalysis activity.

One of the most intriguing possibilities for plasmonic photocatalysts is the selective deposition of energy to specific adsorbate energy states to enable state-selective reactions. Designing plasmonic catalysts with controllable eletronic interaction between excited plasmon and adsorbate molecules is therefore an area that is attracting significant attention in the field. Simultaneous control of the hot carriers and local plasmon heating will open up reaction pathways that are impossilbe with traditional catalysts. In addition, although the thoretical foundation for LSPR has been established, computational studies of plasmon enhanced photocatalysis processes are lagging behind experimental progress. Theoretical work will be an essential element for understanding the structure-function relationship and for enhancing the performance of these photocatalysts. Moreover, recently there has been an increased interest in expanding the plasmonic materials beyond the most widely studied Au and Ag systems to include, for example, Al with UV LSPR and conducting metal oxides whose plasmon resonance can extend deep into the NIR spectral region. In particular, effectively utilizing low energy visible to NIR photons for photocatalysis applications is crucial to realizing the long sought after goal of efficiently using sunlight for photocatalytic processes. Finally, large scale applications of plasmon enhanced photocatalysis should be an intensively studied area in order to realize the many promises of this interesting class of photocatalysts.

5. Acknowledgment

This technical effort was performed under RES contract DE-FE0004000. This report was prepared as an account of work sponsored by an agency of the United States Government. Neither the United States Government nor any agency thereof, nor any of their employees, makes any warranty, express or implied, or assumes any legal liability of responsibility for the accuracy, completeness, or usefulness of any information, apparatus, product, or process disclosed, or represents that its use would not infringe privately owned rights. Reference herein to any specific commercial product, process, or service by trade name, trademark, manufacturer, or otherwise does not necessarily constitute or imply its endorsement, recommendation, or favoring by the United States Government or any agency thereof. The views and opinions of authors expressed herein do not necessarily state or reflect those of the United States Government or any agency thereof.

References

1. V. Myroshnychenko, J. Rodriguez-Fernandez, I. Pastoriza-Santos, A. M. Funston, C. Novo, P. Mulvaney, L. M. Liz-Marzan and F. J. Garcia de Abajo. Modelling the optical response of gold nanoparticles. *Chem. Soc. Rev.* **37** (9), 2008, 1792–1805.
2. (a) M. A. El-Sayed, Some Interesting Properties of Metals Confined in Time and Nanometer Space of Different Shapes. *Accounts of Chemical Research* **34**, 2001, 257–264; (b) K. L. Kelly, E. Coronado, L. L. Zhao and G. C. Schatz. The Optical Properties of Metal Nanoparticles: The Influence of Size, Shape, and Dielectric Envrionment. *J. Phys. Chem. B* **107**, 2003, 668–677.
3. (a) M. W. Knight, N. S. King, L. Liu, H. O. Everitt, P. Nordlander and N. J. Halas. Aluminum for Plasmonics. *ACS Nano* **8** (1), 2014, 834–840; (b) M. Liu and P. Guyot-Sionnest. Mechanism of Silver(I)-Assisted Growth of Gold Nanorods and Bipyramids. *J. Phys. Chem. B* **109** (47), 2005, 22192–22200; (c) B. Wiley, Y. Sun and Y. Xia. Synthesis of Silver Nanostructures with Controlled Shapes and Properties. *Accounts of Chemical Research* **40** (10), 2007, 1067–1076; (d) S. E. Lohse and C. J. Murphy. The Quest for Shape Control: A History of Gold Nanorod Synthesis. *Chem. Mater.* **25** (8),

2013, 1250–1261; (e) H. Chen, L. Shao, Q. Li and J. Wang. Gold nanorods and their plasmonic properties. *Chem. Soc. Rev.* **42**, 2013, 2679–2724; (f) R. Buonsanti, A. Llordes, S. Aloni, B. A. Helms and D. J. Milliron. Tunable Infrared Absorption and Visible Transparency of Colloidal Aluminum-Doped Zinc Oxide Nanocrystals. *Nano Lett.* **11** (11), 2011, 4706–4710; (g) T. R. Gordon, T. Paik, D. R. Klein, G. V. Naik, H. Caglayan, A. Boltasseva and C. B. Murray. Shape-Dependent Plasmonic Response and Directed Self-Assembly in a New Semiconductor Building Block, Indium-Doped Cadmium Oxide (ICO). *Nano Lett.* **13**, 2013, 2857–2863.
4. (a) K. A. Willets and R. P. Van Duyne. Localized Surface Plasmon Resonance Spectroscopy and Sensing. *Ann. Rev. Phys. Chem.* **58** (1), 2007, 267–297; (b) P. Zijlstra, P. M. R. Paulo and M. Orrit. Optical detection of single non-absorbing molecules using the surface plasmon resonance of a gold nanorod. *Nat. Nanotechnol.* **7** (6), 2012, 379–382.
5. H. A. Atwater and A. Polman. Plasmonics for improved photovoltaic devices. *Nat. Mater.* **9** (3), 2010, 205–213.
6. A. Boltasseva and H. A. Atwater. Low-Loss Plasmonic Metamaterials. *Science* **331** (6015), 2011, 290–291.
7. (a) J. A. Schuller, E. S. Barnard, W. Cai, Y. C. Jun, J. S. White and M. L. Brongersma. Plasmonics for extreme light concentration and manipulation. *Nat. Mater.* **9** (3), 2010, 193–204; (b) S. Lal, S. Link and N. J. Halas. Nano-optics from sensing to waveguiding. *Nat. Photon.* **1** (11), 2007, 641–648; (c) M. I. Stockman. Nanofocusing of Optical Energy in Tapered Plasmonic Waveguides. *Phys. Rev. Lett.* **93** (13), 2004, 137404.
8. (a) G. Baffou, C. Girard and R. Quidant. Mapping Heat Origin in Plasmonic Structures. *Physical Review Letters* **104**, 2010, 136805; (b) A. O. Govorov and H. H. Richardson. Generating heat with metal nanoparticles. *Nano Today* **2** (1), 2007, 30–38; (c) O. Neumann, A. S. Urban, J. Day, S. Lal, P. Nordlander and N. J. Halas. Solar Vapor Generation Enabled by Nanoparticles. *ACS Nano* **7**, 2013, 42–49.
9. (a) M. Hu, J. Chen, Z.-Y. Li, L. Au, G. V. Hartland, X. Li, M. Marquez and Y. Xia. Gold nanostructures: engineering their plasmonic properties for biomedical applications. *Chem. Soc. Rev.* **35** (11), 2006, 1084–1094; (b) X. Huang, P. K. Jain, I. H. El-Sayed and M. A. El-Sayed. Plasmonic photothermal therapy (PPTT) using gold nanoparticles. *Lasers Med. Sci.* **23** (3), 2008, 217–228.
10. M. T. Sheldon, J. van de Groep, A. M. Brown, A. Polman and H. A. Atwater. Plasmoelectric potentials in metal nanostructures. *Science* **346** (6211), 2014, 828–831.

11. S. Bai, J. Jiang, Q. Zhang and Y. Xiong. Steering charge kinetics in photocatalysis: intersection of materials syntheses, characterization techniques and theoretical simulations. *Chem. Soc. Rev.* **44** (10), 2015, 2893–2939.
12. (a) I. Thomann, B. A. Pinaud, Z. Chen, B. M. Clemens, T. F. Jaramillo and M. L. Brongersma. Plasmon Enhanced Solar-to-Fuel Energy Conversion. *Nano Lett.* **11** (8), 2011, 3440–3446; (b) H. Gao, C. Liu, H. E. Jeong and P. Yang. Plasmon-Enhanced Photocatalytic Activity of Iron Oxide on Gold Nanopillars. *ACS Nano* **6** (1), 2012, 234–240; (c) C. An, S. Peng and Y. Sun. Facile Synthesis of Sunlight-Driven AgCl:Ag Plasmonic Nanophotocatalyst. *Advanced Materials* **22** (23), 2010, 2570–2574.
13. (a) A. Marimuthu, J. Zhang and S. Linic. Tuning selectivity in propylene epoxidation by plasmon mediated photo-switching of Cu oxidation state. *Science* **339**, 2013, 1590–1593; (b) D. A. Boyd, L. Greengard, M. Brongersma, M. Y. El-Naggar and D. G. Goodwin. Plasmon-Assisted Chemical Vapor Deposition. *Nano Lett.* **6**, 2006, 2592–2597; (c) C. Wang, O. Ranasingha, S. Natesakhawat, P. Ohodinicki, M. Andio, J. P. Lewis and C. Matranga. Visible light plasmonic heating of Au–ZnO for the catalytic reduction of CO_2. *Nanoscale* **5**, 2013, 6968–6974.
14. (a) W. Hou and S. B. Cronin. A Review of Surface Plasmon Resonance-Enhanced Photocatalysis. *Adv. Funct. Mater.* **23**, 2012, 1612–1619; (b) C. Clavero. Plasmon-induced hot-electron generation at nanoparticle/metal-oxide interfaces for photovoltaic and photocatalytic devices. *Nat. Photon.* **8** (2), 2014, 95–103; (c) M. J. Kale, T. Avanesian and P. Christopher. Direct Photocatalysis by Plasmonic Nanostructures. *ACS Catal.* **4**, 2014, 116–128; (d) M. L. Brongersma, N. J. Halas and P. Nordlander. Plasmon-induced hot carrier science and technology. *Nat. Nanotechnol.* **10** (1), 2015, 25–34; (e) S. Linic, U. Aslam, C. Boerigter and M. Morabito. Photochemical transformations on plasmonic metal nanoparticles. *Nat. Mater.* **14** (6), 2015, 567–576.
15. A. Nitzan and L. E. Brus. Theoretical model for enhanced photochemistry on rough surfaces. *J. Chem. Phys.* **75** (5), 1981, 2205–2214.
16. C. J. Chen and R. M. Osgood. Direct Observation of the Local-Field-Enhanced Surface Photochemical Reactions. *Phys. Rev. Lett.* **50** (21), 1983, 1705–1708.
17. P. K. Jain, K. S. Lee, I. H. El-Sayed, and M. A. El-Sayed. Calculated Absorption and Scattering Properties of Gold Nanoparticles of Different Size, Shape, and Composition: Applications in Biological Imaging and Biomedicine. *J. Phys. Chem. B* **110** (14), 2006, 7238–7248.
18. X. Lu, M. Rycenga, S. E. Skrabalak, B. Wiley and Y. Xia. Chemical Synthesis of Novel Plasmonic Nanoparticles. *Ann. Rev. Phys. Chem.* **60**, 2009, 167–192.

19. X. Chen, H.-Y. Zhu, J.-C. Zhao, Z.-F. Zheng and X.-P. Gao. Visible-Light-Driven Oxidation of Organic Contaminants in Air with Gold Nanoparticle Catalysts on Oxide Supports. *Angew. Chem. Int. Edit.* **47**, 2008, 5353–5356.
20. (a) O. K. Ranasingha, C. Wang, P. R. Ohodnicki, J. W. Lekse, J. P. Lewis and C. Matranga. Synthesis, characterization, and photocatalytic activity of Au-ZnO nanopyramids. *J. Mater. Chem. A* **3** (29), 2015, 15141–15147; (b) S. Navalon, de M. Miguel, R. Martin, M. Alvaro and Garcia, H. Enhancement of the Catalytic Activity of Supported Gold Nanoparticles for the Fenton Reaction by Light. *J. Am. Chem. Soc.* **133** (7), 2011, 2218–2226.
21. F. Wang, C. Li, H. Chen, R. Jiang, L.-D. Sun, Q. Li, J. Wang, J. C. Yu and C.-H. Yan. Plasmonic Harvesting of Light Energy for Suzuki Coupling Reactions. *J. Am. Chem. Soc.* **135** (15), 2013, 5588–5601.
22. (a) Z. Liu, W. Hou, P. Pavaskar, M. Aykol and S. B. Cronin. Plasmon Resonant Enhancement of Photocatalytic Water Splitting Under Visible Illumination. *Nano Lett.* **11**, 2011, 1111–1116; (b) J. Lee, S. Mubeen, X. L. Ji, G. D. Stucky and M. Moskovits. Plasmonic Photoanodes for Solar Water Splitting with Visible Light. *Nano Lett.* **12** (9), 2012, 5014–5019.
23. W. Hou, W. H. Hung, P. Pavaskar, A. Goeppert, M. Aykol and S. B. Cronin. Photocatalytic Conversion of CO_2 to Hydrocarbon Fuels via Plasmon-Enhanced Absorption and Metallic Interband Transitions. *ACS Catal.* **1**, 2011, 929–936.
24. (a) G. V. Hartland. Optical Studies of Dynamics in Noble Metal Nanostructures. *Chem. Rev.* **111** (6), 2011, 3858–3887; (b) A. Manjavacas, J. G. Liu, V. Kulkarni and P. Nordlander. Plasmon-Induced Hot Carriers in Metallic Nanoparticles. *ACS Nano* **8** (8), 2014, 7630–7638.
25. C. F. Bohren. How can a particle absorb more than the light incident on it? *Am. J. Phys.* **51** (4), 1983, 323–327.
26. K. R. Catchpole and A. Polman. Plasmonic solar cells. *Opt. Express* **16** (26), 2008, 21793–21800.
27. E. Hao and G. C. Schatz. Electromagnetic fields around silver nanoparticles and dimers. *J. Chem. Phys.* **120** (1), 2004, 357–366.
28. S. L. Kleinman, R. R. Frontiera, A.-I. Henry, J. A. Dieringer and R. P. Van Duyne. Creating, characterizing, and controlling chemistry with SERS hot spots. *Phys. Chem. Chem. Phys.* **15** (1), 2013, 21–36.
29. D.-Y. Wu, L.-B. Zhao, X.-M. Liu, R. Huang, Y.-F. Huang, B. Ren and Z.-Q. Tian. Photon-driven charge transfer and photocatalysis of p-aminothiophenol in metal nanogaps: a DFT study of SERS. *Chem. Commun.* **47** (9), 2011, 2520–2522.

30. W. Hou, Z. Liu, P. Pavaskar, W. H. Hung and S. B. Cronin. Plasmonic enhancement of photocatalytic decomposition of methyl orange under visible light. *J.Catal.* **277**, 2011, 149–153.
31. W. H. Hung, M. Aykol, D. Valley, W. Hou and S. B. Cronin. Plasmon Resonant Enhancement of Carbon Monoxide Catalysis. *Nano Lett.* **10**, 2010, 1314–1318.
32. (a) A. Kubacka, M. Fernández-García and G. Colón. Advanced Nanoarchitectures for Solar Photocatalytic Applications. *Chem. Rev.* **112** (3), 2012, 1555–1614; (b) D. M. Schultz and T. P. Yoon. Solar Synthesis: Prospects in Visible Light Photocatalysis. *Science* **343** (6174), 2014, 1239176.
33. H. Zhu, X. Ke, X. Yang, S. Sarina and H. Liu. Reduction of Nitroaromatic Compounds on Supported Gold Nanoparticles by Visible and Ultraviolet Light. *Angew. Chem. Int. Edit.* **49** (50), 2010, 9657–9661.
34. H. Li, Z. Bian, J. Zhu, Y. Huo, H. Li and Y. Lu. Mesoporous Au/TiO_2 Nanocomposites with Enhanced Photocatalytic Activity. *J. Am. Chem. Soc.* **129** (15), 2007, 4538–4539.
35. D. B. Ingram, P. Christopher, J. L. Bauer and S. Linic. Predictive Model for the Design of Plasmonic Metal/Semiconductor Composite Photocatalysts. *ACS Catal.* **1** (10), 2011, 1441–1447.
36. T. Torimoto, H. Horibe, T. Kameyama, K.-I. Okazaki, S. Ikeda, M. Matsumura, A. Ishikawa and H. Ishihara. Plasmon-Enhanced Photocatalytic Activity of Cadmium Sulfide Nanoparticle Immobilized on Silica-Coated Gold Particles. *J. Phys. Chem. Lett.* **2** (16), 2011, 2057–2062.
37. A. Takai and P. V. Kamat. Capture, Store, and Discharge. Shuttling Photogenerated Electrons across TiO_2–Silver Interface. *ACS Nano* **5** (9), 2011, 7369–7376.
38. (a) A. Wood, M. Giersig and P. Mulvaney. Fermi Level Equilibration in Quantum Dot–Metal Nanojunctions. *The Journal of Physical Chemistry B* **105** (37), 2001, 8810–8815; (b) V. Subramanian, E. E. Wolf, P. V. Kamat. Catalysis with TiO_2/Gold Nanocomposites. Effect of Metal Particle Size on the Fermi Level Equilibration. *J. Am. Chem. Soc.* **126** (15), 2004, 4943–4950.
39. (a) R. K. Quinn, R. D. Nasby and R. J. Baughman, Photoassisted electrolysis of water using single crystal α-Fe_2O_3 anodes. *Mater. Res. Bull.* **11** (8), 1976, 1011–1017; (b) J. H. Kennedy and K. W. Frese. Photooxidation of Water at α-Fe_2O_3 Electrodes. *J. Electrochem. Soc.* **125** (5), 1978, 709–714.
40. S. Linic, P. Christopher and D. B. Ingram. Plasmonic-metal nanostructures for efficient conversion of solar to chemical energy. *Nat. Mater.* **10**, 2011, 911–921.

41. G. Zhao, H. Kozuka and T. Yoko. Sol-gel preparation and photoelectrochemical properties of TiO_2 films containing Au and Ag metal particles. *Thin Sol. Films* **277** (1–2), 1996, 147–154.
42. (a) T. Lana-Villarreal and R. Gómez. Tuning the photoelectrochemistry of nanoporous anatase electrodes by modification with gold nanoparticles: Development of cathodic photocurrents. *Chem. Phys. Lett.* **414** (4–6), 2005, 489–494; (b) K. Yu, Y. Tian and T. Tatsuma. Size effects of gold nanoparticles on plasmon-induced photocurrents of gold-TiO_2 nanocomposites. *Phys. Chem. Chem. Phys.* **8** (46), 2006, 5417–5420; (c) N. Sakai, Y. Fujiwara, Y. Takahashi and T. Tatsuma. Plasmon-Resonance-Based Generation of Cathodic Photocurrent at Electrodeposited Gold Nanoparticles Coated with TiO_2 Films. *Chem. Phys. Chem.* **10** (5), 2009, 766–769; (d) E. Kowalska, O. O. P. Mahaney, R. Abe and B. Ohtani. Visible-light-induced photocatalysis through surface plasmon excitation of gold on titania surfaces. *Phys. Chem. Chem. Phys.* **12** (10), 2010, 2344–2355; (e) C. Gomes Silva, R. Juárez, T. Marino, R. Molinari and H. García. Influence of Excitation Wavelength (UV or Visible Light) on the Photocatalytic Activity of Titania Containing Gold Nanoparticles for the Generation of Hydrogen or Oxygen from Water. *J. Am. Chem. Soc.* **133** (3), 2011, 595–602; (f) Y. Ide, M. Matsuoka and M. Ogawa. Efficient Visible-Light-Induced Photocatalytic Activity on Gold-Nanoparticle-Supported Layered Titanate. *J. Am. Chem. Soc.* **132** (47), 2010, 16762–16764; (g) A. Tanaka, A. Ogino, M. Iwaki, K. Hashimoto, A. Ohnuma, F. Amano, B. Ohtani and H. Kominami. Gold–Titanium(IV) Oxide Plasmonic Photocatalysts Prepared by a Colloid-Photodeposition Method: Correlation Between Physical Properties and Photocatalytic Activities. *Langmuir* **28** (36), 2012, 13105–13111; (h) X. Shi, K. Ueno, N. Takabayashi and H. Misawa. Plasmon-Enhanced Photocurrent Generation and Water Oxidation with a Gold Nanoisland-Loaded Titanium Dioxide Photoelectrode. *J. Phys. Chem. C* **117** (6), 2013, 2494–2499; (i) N. Sakai, T. Sasaki, K. Matsubara and T. Tatsuma. Layer-by-layer assembly of gold nanoparticles with titania nanosheets: control of plasmon resonance and photovoltaic properties. *J. Mater. Chem.* **20** (21), 2010, 4371–4378.
43. (a) Z. H. Chen, Y. B. Tang, C. P. Liu, Y. H. Leung, G. D. Yuan, L. M. Chen, Y. Q. Wang, I. Bello, J. A. Zapien, W. J. Zhang, C. S. Lee and S. T. Lee. Vertically Aligned ZnO Nanorod Arrays Sentisized with Gold Nanoparticles for Schottky Barrier Photovoltaic Cells. *J. Phys. Chem. C* **113** (30), 2009, 13433–13437; (b) H. M. Chen, C. K. Chen, C.-J. Chen, L.-C. Cheng, P. C. Wu, B. H. Cheng, Y. Z. Ho, M. L. Tseng, Y.-Y. Hsu, T.-S. Chan, J.-F. Lee, R.-S. Liu and D. P. Tsai. Plasmon Inducing Effects for Enhanced

Photoelectrochemical Water Splitting: X-ray Absorption Approach to Electronic Structures. *ACS Nano* **6** (8), 2012, 7362–7372.
44. (a) A. Primo, T. Marino, A. Corma, R. Molinari and H. García. Efficient Visible-Light Photocatalytic Water Splitting by Minute Amounts of Gold Supported on Nanoparticulate CeO_2 Obtained by a Biopolymer Templating Method. *J. Am. Chem. Soc.* **133** (18), 2011, 6930–6933; (b) H. Kominami, A. Tanaka and K. Hashimoto. Mineralization of organic acids in aqueous suspensions of gold nanoparticles supported on cerium(iv) oxide powder under visible light irradiation. *Chem. Commun.* **46** (8), 2010, 1287–1289.
45. X. Zhou, C. Hu, X. Hu, T. Peng and J. Qu. Plasmon-Assisted Degradation of Toxic Pollutants with $Ag–AgBr/Al_2O_3$ under Visible-Light Irradiation. *J. Phys. Chem. C* **114** (6), 2010, 2746–2750.
46. P. Wang, B. Huang, X. Qin, X. Zhang, Y. Dai, J. Wei and M.-H. Whangbo. Ag@AgCl: A Highly Efficient and Stable Photocatalyst Active under Visible Light. *Angew. Chem. Int. Edit.* **47** (41), 2008, 7931–7933.
47. C. Hu, T. Peng, X. Hu, Y. Nie, X. Zhou, J. Qu and H. He. Plasmon-Induced Photodegradation of Toxic Pollutants with $Ag–AgI/Al_2O_3$ under Visible-Light Irradiation. *J. Am. Chem. Soc.* **132** (2), 2010, 857–862.
48. (a) Z. Zheng, B. Huang, X. Qin, X. Zhang, Y. Dai and M.-H. Whangbo. Facile in situ synthesis of visible-light plasmonic photocatalysts M@TiO2 (M = Au, Pt, Ag) and evaluation of their photocatalytic oxidation of benzene to phenol. *J. Mater. Chem.* **21** (25), 2011, 9079–9087; (b) N. Zhang, S. Liu, X. Fu and Y.-J. Xu. Synthesis of $M@TiO_2$ (M = Au, Pd, Pt) Core–Shell Nanocomposites with Tunable Photoreactivity. *J. Phys. Chem.* C **115** (18), 2011, 9136–9145.
49. (a) Y. Tian, and T. Tatsuma. Mechanisms and Applications of Plasmon-Induced Charge Separation at TiO_2 Films Loaded with Gold Nanoparticles. *Journal of the American Chemical Society* **127** (20), 2005, 7632–7637; (b) A. Furube, L. Du, K. Hara, R. Katoh and M. Tachiya. Ultrafast Plasmon-Induced Electron Transfer from Gold Nanodots into TiO_2 Nanoparticles. *Journal of the Am. Chem. Soc.* **129** (48), 2007, 14852–14853.
50. K. Wu, W.-E. Rodríguez-Córdoba, Y. Yang, and T. Lian. Plasmon-Induced Hot Electron Transfer from the Au Tip to CdS Rod in CdS-Au Nanoheterostructures. *Nano Lett.* **13** (11), 2013, 5255–5263.
51. (a) S. Mubeen, J. Lee, N. Singh, S. Kramer, G. D. Stucky and M. Moskovits. An autonomous photosynthetic device in which all charge carriers derive from surface plasmons. *Nature Nanotechnology* 2013, **8** (4), 247–251; (b) Y. Zhong, K. Ueno, Y. Mori, X. Shi, T. Oshikiri, K. Murakoshi, H. Inoue and H. Misawa. Plasmon-Assisted Water Splitting Using Two Sides of the

Same SrTiO$_3$ Single-Crystal Substrate: Conversion of Visible Light to Chemical Energy. *Angew. Chem. Int. Edit.* **53** (39), 2014, 10350–10354.
52. (a) H. Petek and S. Ogawa. Femtosecond time-resolved two-photon photoemission studies of electron dynamics in metals. *Prog. Surf. Sci.* **56** (4), 1997, 239–310; (b) M. Bonn, S. Funk, C. Hess, D. N. Denzler, C. Stampfl, M. Scheffler, M. Wolf and G. Ertl. Phonon- Versus Electron-Mediated Desorption and Oxidation of CO on Ru(0001). *Science* **285** (5430), 1999, 1042–1045; (c) C. Frischkorn and M. Wolf. Femtochemistry at Metal Surfaces: Nonadiabatic Reaction Dynamics. *Chem. Rev.* **106** (10), 2006, 4207–4233.
53. J. Gavnholt, A. Rubio, T. Olsen, K. S. Thygesen and J. Schiøtz. Hot-electron-assisted femtochemistry at surfaces: A time-dependent density functional theory approach. *Phys. Rev. B* **79** (19), 2009, 195405.
54. K. Watanabe, D. Menzel, N. Nilius and H.-J. Freund. Photochemistry on Metal Nanoparticles. *Chem. Rev.* **106** (10), 2006, 4301–4320.
55. S. Mukherjee, L. Zhou, A. M. Goodman, N. Large, C. Ayala-Orozco, Y. Zhang, P. Nordlander and N. J. Halas. Hot-Electron-Induced Dissociation of H$_2$ on Gold Nanoparticles Supported on SiO$_2$. *J. Am. Chem. Soc.* **136** (1), 2014, 64–67.
56. S. Mukherjee, F. Libisch, N. Large, O. Neumann, L. V. Brown, J. Cheng, J. B. Lassiter, E. A. Carter, P. Nordlander and N. J. Halas. Hot Electrons Do the Impossible: Plasmon-Induced Dissociation of H$_2$ on Au. *Nano Lett.* **13** (1), 2013, 240–247.
57. P. Christopher, H. Xin and S. Linic. Visible-light-enhanced catalytic oxidation reactions on plasmonic silver nanostructures. *Nat. Chem.* **3**, 2011, 467–472.
58. P. Christopher, H. Xin, A. Marimuthu and S. Linic. Singular characteristics and unique chemical bond activation mechanisms of photocatalytic reactions on plasmonic nanostructures. *Nat. Mater.* **11**, 2012, 1044–1050.
59. T. E. Madey, J. T. Yates, D. A. King and C. J. Uhlaner. Isotope Effect in Electron Stimulated Desorption: Oxygen Chemisorbed on Tungsten. *J. Chem. Phys.* **52** (10), 1970, 5215–5220.
60. Z. Bian, T. Tachikawa, P. Zhang, M. Fujitsuka and T. Majima. Au/TiO$_2$ Superstructure-Based Plasmonic Photocatalysts Exhibiting Efficient Charge Separation and Unprecedented Activity. *J. Am. Chem. Soc.* **136** (1), 2014, 458–465.
61. A. O. Govorov, H. Zhang and Y. K. Gun'ko. Theory of Photoinjection of Hot Plasmonic Carriers from Metal Nanostructures into Semiconductors and Surface Molecules. *J. Phys. Chem.* C **117**, 2013, 16616–16631.

62. A. Tanaka, S. Sakaguchi, K. Hashimoto and H. Kominami. Preparation of Au/TiO$_2$ with Metal Cocatalysts Exhibiting Strong Surface Plasmon Resonance Effective for Photoinduced Hydrogen Formation under Irradiation of Visible Light. *ACS Catal.* **3** (1), 2013, 79–85.
63. L. Cao, D. N. Barsic, A. R. Guichard and M. L. Brongersma. Plasmon-Assisted Local Temperature Control to Pattern Individual Semiconductor Nanowires and Carbon Nanotubes. *Nano Lett.* **7**, 2007, 3523–3527.
64. W. H. Hung, I. K. Hsu, A. Bushmaker, R. Kumar, J. Theiss and S. B. Cronin. Laser Directed Growth of Carbon-Based Nanostructures by Plasmon Resonant Chemical Vapor Deposition. *Nano Lett.* **8** (10), 2008, 3278–3282.
65. C.-W. Yen and M. A. El-Sayed. Plasmonic Field Effect on the Hexacyanoferrate (III)-Thiosulfate Electron Transfer Catalytic Reaction on Gold Nanoparticles: Electromagnetic or Thermal? *J. Phys. Chem.* C **113** (45), 2009, 19585–19590.
66. J. R. Adleman, D. A. Boyd, D. G. Goodwin and D. Psaltis. Heterogenous Catalysis Mediated by Plasmon Heating. *Nano Lett.* **9** (12), 2009, 4417–4423.

Chapter 6

Plasmon-assisted Chemical Reactions

Ruibin Jiang and Jianfang Wang[*]

Department of Physics
The Chinese University of Hong Kong
Shatin, Hong Kong SAR, PRC
[*]jfwang@phy.cuhk.edu.hk

The transformation of starting organic materials to valuable target compounds in a green, effective and low-cost manner is always pursued, especially under ambient conditions. Plasmonic catalysts provide an opportunity to the realization of the transformation of organic compounds under visible-light illumination at ambient temperatures. In this Chapter, we introduce the applications of plasmonic catalysts in organic synthesis. The concept of localized surface plasmon resonance and its characteristics as well as applications are first introduced. Then, the functioning mechanisms of plasmon in photocatalysis are elucidated, followed by the reported applications of plasmonic catalysis in organic synthesis. Finally, the future developments of plasmonic catalysis are discussed.

1. Introduction

The development of the human society is highly dependent on fossil fuels. Due to the limited reserves of fossil fuels and environmental pollution from the combustion of fossil fuels, global energy and environmental issues have been brought to the attention of chemists and technologists in the 21st century. Many efforts have been devoted to the development of technologies that can create sustainable and environmentally benign energy or can save the energy consumptions in industries and daily life. Solar energy has been widely accepted as a free, abundant and endlessly renewable source of clean energy, which can meet current and future human energy demands. The harvest and conversion of solar energy into usable energy forms or the direct use of solar energy in industries is highly desirable.[1,2] To date, there have been three primary solar energy technologies, including solar heating, solar photovoltaics and solar thermal electricity.[3,4] In the first technology, solar energy is used to heat water. In the second and third technologies, solar energy is converted into electricity. The converted electricity must be used immediately or stored in a secondary device, for example, capacitors and batteries. The exploration of new solar energy utilization technologies is therefore of importance and will have far-reaching significance.

Photocatalysis is a very promising solar energy harvesting technology that can directly produce chemical fuels or can directly use solar energy to drive chemical reactions in chemical industries.[5-7] Chemical industries consume a large fraction of energy in heating chemical reactions. The use of solar energy to drive chemical reactions can therefore save a tremendous amount of energy required to heat reactions in chemical industries. However, to date, the energy conversion efficiency in photocatalysis is not high, which is mainly limited by photocatalysts. Most photocatalysts are based on semiconductors. When semiconductors are excited under light with photon energies larger than the band gaps, electron–hole pairs are generated in semiconductors. The generated high energy electrons/holes can reduce/oxidize chemicals. Unfortunately, most of stable semiconductors, such as TiO_2, ZnO, CeO_2 have wide bandgaps. They cannot be excited or are excited only weakly by visible and near-infrared (NIR) light, which accounts for approximately a half of solar energy. Many methods, such as doping with metal or non-metal elements, tailoring the morphology and forming heterostructures, have been employed to improve the photoresponse of semiconductors in the visible and NIR regions.[8] Besides semiconductors, plasmonic metal nanostructures have recently been found to show photocatalytic activities under resonant excitation.[9-15] Because the localized surface plasmon resonances (LSPRs) of metal nanostructures can be readily controlled from the visible to NIR region by varying the morphology of the nanostructures, plasmonic nanostructures are promising photocatalysts to the achievement of high energy conversion efficiencies. Plasmonic photocatalysts have been used in water splitting,[16-18] carbon dioxide reduction,[12,19] degradation of organic pollutant and organic synthesis.[10,13]

In this chapter, we will introduce the application of plasmonic photocatalysts in organic synthesis, which can save the energy consumed in chemical industries. The properties of localized surface plasmon will be introduced first, followed by the discussion of the functioning mechanisms of plasmon in chemical reactions. Plasmon-assisted organic synthesis will then be reviewed according to the types of reactions. Finally, we will give our outlooks and concluding remarks for the future developments of plasmonic photocatalysts in organic synthesis.

2. Properties of Localized Surface Plasmon Resonance

A localized surface plasmon is a collective oscillation of the free electrons in a highly conductive nanostructure. It can be described as a quantum of plasma oscillation. It can also be described accurately by classical electrodynamics. In classical physics, one can think of these plasmon oscillations as mechanical oscillations of the electron gas in a nanostructure. The oscillations are stimulated by the oscillating electric field of a light wave, where the electric field causes displacements of the electron gas with respect to the fixed ionic cores. A given nanostructure in a fixed environment usually has several inherent plasmon frequencies. When the external excitation electric field arising from light or electrons have the same frequency as the inherent plasmon frequency, plasmon resonance occurs, that is, the collective oscillation amplitude reaches a maximum. Upon resonant excitation, the optical response, including the absorption and scattering cross-sections, of the nanostructure and the local electromagnetic field in the nanoscale spatial region around the nanostructure are remarkably enhanced. It can be readily deduced from classical physics that the LSPR frequencies, optical cross-sections and electromagnetic field enhancements of a nanostructure are dependent on the material, size, shape and crystalline structure of the nanostructure, as well as its surrounding environment.[20,21] In order to facilitate researchers from chemistry and materials sciences to understand these dependences, we will elucidate them using a simple physical picture instead of going through complicated formula derivations, the latter of which can be found in some books and review papers.[20,22,23] The dependence on materials stems from the fact that the number of free electrons per volume and the mobility of free electrons are both determined by the material type. Materials that can support LSPRs in the visible and NIR regions are required to have free charge carrier concentrations up to 10^{21} cm^{-3}.[24] As a result, metals, such as Au, Ag and Cu, are good materials for supporting LSPRs in the visible and near-infrared regions. Both the size and shape of nanostructures can affect the restoring force between free electrons and the fixed ionic cores as well as the electron distribution on the nanostructure surface and therefore influence the LSPR properties. Defects in crystalline nanostructures can scatter electrons and thereby destroy the oscillation coherence. The dependence on the surrounding environment can be understood by the fact that plasmon resonance can induce polarized charges around the nanostructure and the polarized charges in turn exert an extra force on oscillation electrons through electrostatic interaction, which alters the plasmon resonance energy. Different surrounding materials have distinct capabilities in generating polarized charges and therefore affect the plasmon resonance differently.

Under resonant excitation, plasmonic metal nanocrystals interact strongly with light. In other words, LSPR endows metal nanocrystals with very large absorption and scattering cross-sections at the LSPR wavelength. Compared with other optical species, including atoms/ions, organic dyes and semiconductor and carbon dots, plasmonic metal nanocrystals have optical cross-sections larger than their physical cross-sections. Moreover, both of their absorption and scattering cross-sections are important (Fig. 1(a)).[25] Another prominent property of plasmonic metal nanocrystals is the electric field localization. Upon resonant excitation, LSPR can concentrate light beyond the diffraction limit into a nanoscale region.[26,27] The strong light localization can make the optical electric field near the surface of plasmonic metal nanocrystals several orders of magnitude larger than the incident field (Fig. 1(b)). Another attractive feature of LSPR is its facile spectral controllability. For a given material, the LSPR wavelength can be readily tailored by the nanocrystal shape and size. Taking Au nanorods as an example, the wavelength of the longitudinal plasmon resonance, which corresponds to the collective electron oscillations along the nanorod length axis, can be varied from the visible to NIR region by only changing the length/diameter ratio (Fig. 1(c)). Because of their attractive optical properties, plasmonic metal nanocrystals have been widely used in various fields from nanophotonics, biotechnology, analytical chemistry and photocatalysis to information storage.[21]

After plasmon resonances are excited, they are damped rapidly, on the order of femtoseconds, through radiative and non-radiative channels

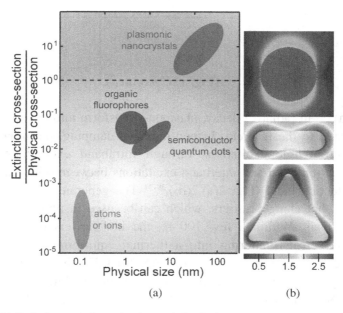

Fig. 1. (a) Ratio between the extinction and physical cross-sections vs. the physical size of optical species. (b) Logarithmic scale electric field intensity enhancement contours of differently shaped Au nanocrystals with the same volume. The Au nanosphere (top) has a diameter of 50 nm. The length and diameter of the Au nanorod (middle) are 102 nm and 30 nm, respectively. The Au nanoplate (bottom) in an equilaterally triangular shape has an edge length of 87 nm and a thickness of 10 nm. (c) Extinction spectra of Au nanorods with different length-to-diameter aspect ratios. The longitudinal plasmon resonance wavelength can be tailored from the visible to NIR region.

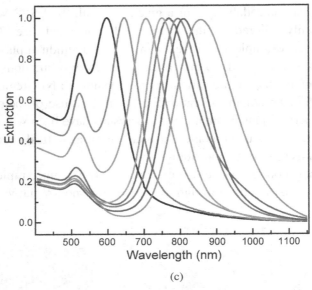

(c)

(Fig. 1. *Continued*)

(Fig. 2(a)).[28] In radiative decay, plasmons transform into photons, which are radiated to the far-field. In non-radiative damping, plasmons decay into electron–hole excitations through intraband excitations within the conduction band and interband excitations between other bands and the conduction band (Fig. 2(a),2(b)).[29–31] The generated electrons/holes are called hot charge carriers, which can be used in photovoltaics and photocatalysis. Within a few 100 fs, the hot carriers decay through electron–electron scattering into a thermal equilibrium electron gas, which has a Fermi–Dirac distribution (Fig. 2(b)). The thermal equilibrium electron gas further transfers energy to the lattice within a few picoseconds via electron–phonon coupling. The lattice energy is finally transmitted to the surrounding medium through phonon–phonon coupling in a temporal scale of several ten to several hundred picoseconds.

The damping of plasmons is harmful to the quality factor of plasmon resonances. A direct consequence is that the larger the damping, the smaller the electromagnetic field enhancement. However, many applications based on plasmon resonances often benefit from plasmon damping. For example, the hot electrons created by non-radiative damping can be utilized in photocatalysis and photovoltaics.[21] The thermal energy generated by

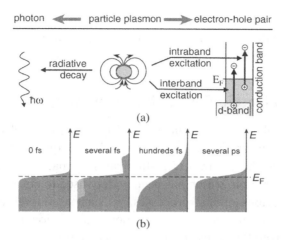

Fig. 2. (a) Schematic representation of radiative (left) and non-radiative (right) decay of plasmons in noble metal nanoparticles. (b) Electron distributions at different time points during plasmon decay. (a) Reproduced with permission.[28]

non-radiative damping can be employed to kill cancer cells.[21] The radiative damping can be used to control the emission direction and polarization of fluorescence.[25,26] Clearly, different applications require plasmonic nanostructures to have different damping processes. Although both radiative and non-radiative damping channels coexist at all time, they can be maximized or minimized by controlling the size and shape of nanostructures. In general, with the increase of the nanoparticle size, both radiative and non-radiative damping increase, with the radiative damping increasing more rapidly than the non-radiative damping. For relatively large nanoparticles, the plasmon damping is usually dominated by the radiative channel, while the non-radiative damping dominates the plasmon damping in small nanoparticles. Moreover, plasmons with resonance energies smaller than the interband transition energy show smaller non-radiative damping that those with resonance energies higher than the interband transition energy.

3. Functioning Mechanisms of Plasmons in Chemical Reactions

Several different mechanisms have been proposed in plasmon-assisted chemical reactions, including local electric field enhancement, hot-electron effect and photothermal conversion. The first is derived from plasmon resonances, while the second and third originate from the non-radiative damping of plasmon resonances.

3.1 Local electric field enhancement

As mentioned above, LSPR can remarkably increase the electromagnetic field near the surface of nanostructures. Such large electric fields can greatly enhance the electron transition rates of materials located in the field-enhanced region, which has been widely used to enhance the excitation rate of various optical processes,[21,25] such as fluorescence, Raman scattering, up conversion and high-harmonic generation. One should note that a prerequisite for the use of the focused electric field to enhance electron transition is that the electron transition should spectrally overlap with the plasmon resonance. When a semiconductor is integrated with a plasmonic metal nanocrystal and the absorption of the semiconductor is spectrally overlapped with the plasmon resonance of the metal nanocrystal, an incident light with an appropriate photon energy can simultaneously excite the LSPR in the metal nanocrystal and electrons from the valence-band to the conduction-band in the semiconductor. The excited LSPR generates a very large electric field that can greatly promote the excitation in the part of the semiconductor that is in the vicinity of the metal nanocrystal. The enhanced excitation of the semiconductor means that the number of the electron–hole pairs generated in this part of the semiconductor is increased (Fig. 3(a)). The subsequent photocatalytic processes are therefore enhanced.

The concentrated electric field of the plasmon resonance can also interact with the dipole moment of polar molecules adsorbed on the metal surface, resulting in the activation of polar bonds in molecules (Fig. 3(b)).[32] Moreover, when plasmonic nanocrystals are loaded on zeolites, the very strong local electric field generated by plasmon can enhance the polarized electrostatic field in zeolites.[33] The enhanced polarized electrostatic field in zeolites has the power to polarize molecules adsorbed on the surface or confined in the porous matrix in zeolites and therefore can reduce the energy required for electron transfer or the activation of reactant molecules. This functioning mechanism only works for polar reactant

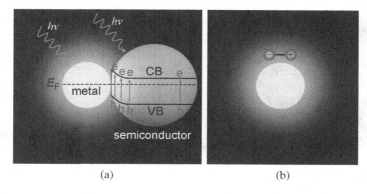

Fig. 3. (a) Plasmon-enhanced excitation of semiconductors. (b) Plasmon-enhanced activation of polar molecules.

molecules. The larger the polarity of the reactant molecules, the higher its photocatalytic activity.

3.2 Hot-electron effect

As mentioned above, plasmon decay can produce high-energy electrons at two stages. The first stage starts from the decay of plasmon into electron–hole pairs and ends before the electrons reach thermal equilibrium. This stage is very short and lasts for less than ~100 fs.[30] The second stage is from the thermalized electrons to the thermal equilibrium reached between the electrons and phonons. The second stage is relatively long and lasts for about several picoseconds.[30] The hot electrons generated in plasmon decay have the possibility to transfer to the conduction-band of semiconductors and to the LUMO of molecules and then accelerate chemical reactions (Fig. 4).

Considering that most oxide semiconductors employed in photocatalysis are n-type, we will use an n-type semiconductor to show the hot-electron transfer between plasmonic metals and semiconductors. When a plasmonic metal is integrated with a n-type semiconductor, the thermal equilibration of the Fermi level gives rise to a potential barrier, the Schottky barrier, between the metal and the semiconductor. The Schottky barrier blocks the electron transfer between the metal and the semiconductor. However, under resonant excitation of the LSPR of the metal

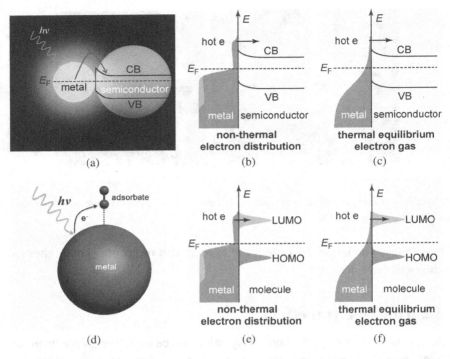

Fig. 4. (a) Schematic of hot-electron injection from a plasmonic nanocrystal to the conduction band of a semiconductor. (b) Non-thermalized hot-electron injection. (c) Thermal-equilibrium hot-electron injection. (d) Schematic of hot-electron transfer from a plasmonic nanocrystal to a molecule. (e) Non-thermalized hot-electron transfer from the plasmonic metal to the LUMO of adsorbed molecules. (f) Thermal-equilibrium hot-electron transfer from the plasmonic metal to the LUMO of adsorbed molecules.

nanocrystal, the fraction of the hot electrons generated during plasmon decay can have energies higher than the Schottky barrier and therefore can cross the barrier and inject into the conduction-band of the semiconductor (Fig. 4(a)). The hot-electron injection can occur at both the stages. The hot-electron injection efficiency from the metal to the semiconductor is determined by the energy and lifetime of the hot electrons and the height of the Schottky barrier. For a given metal semiconductor system, the Schottky barrier is fixed, which is determined by the work function of the metal and the Fermi level of the semiconductor. In the first stage, the number of electrons with energies higher than the Schottky barrier is large, while the lifetime of the electrons is very short (Fig. 4(b)). In contrast, in the second stage, the number of electrons with energies higher than the Schottky barrier is small, while the lifetime of the electrons is relative long (Fig. 4(c)). So far there have been difficulties in telling at which stage the electron injection efficiency is higher. The hot-electron injection confers the photocatalytic reduction activity on the wide bandgap semiconductor that is originally inactive under visible-light.[10,13] On the other hand, the holes left in the metal nanocrystal have a mild oxidation ability, which can be utilized for chemo-selective oxidation of organic molecules.[10,13]

Hot electrons generated by plasmon decay can also directly inject from plasmonic metals to the LUMOs of adsorbed molecules (Fig. 4(d)). Similar to the hot-electron injection in the conduction-band of a semiconductor, the hot-electron transfer from plasmonic metals to adsorbed molecules can also occur at the two stages (Fig. 4(e), 4(f)). The injection of hot electrons into the LUMOs of molecules causes the formation of transient anions.[34,35] In the transient anions, some original chemical bonds are weakened and stretched in order to lower the energy of the transient anions.[34] Subsequently, the injected electron transfers back to the metal nanocrystal, while leaving energy in the bond vibration of the molecule. The average time for the dissipation of the molecular vibrational energy onto the metal surface is on the order of picoseconds, which is long enough to allow for typical chemical transformations.[36] If the transferred energy is sufficient to overcome the activation barrier required for breaking the related bond of molecules, chemical transformations can then take place.[34] In addition, electron transfer in the opposite direction can also occur if a hot hole in the metal nanocrystal attacks an occupied molecular orbital.[37–39] Much still remains to be understood about the use of plasmon-induced hot holes in accelerating chemical reactions.

3.3 Photothermal conversion

If there are no channels for hot-electron transfer to occur, the plasmon resonance energy will be finally converted into thermal energy in the non-radiative decay.[40,41] The ability of a nanoparticle converting attenuated light into thermal energy is usually described by the photothermal conversion efficiency, which is the ratio between the absorption and extinction cross-sections. The photothermal conversion efficiency is dependent on the size and shape of plasmonic nanoparticles. The smaller the nanoparticle size is, the higher its photothermal conversion efficiencis. For the effective diameter of a nanoparticle smaller than ~10 nm, the photothermal conversion efficiency is nearly 100%.[42] Since the absorption of metal nanoparticles originates from the Joule heating effect, the plasmonic photothermal conversion depends on the electric field intensity within the nanoparticle. A nanoparticle with a shape that can effectively screen the electric field therefore has a lower photothermal conversion efficiency in comparison with nanoparticles in other shapes with the same volume.[43] Since the distribution of the electric field within a nanoparticle is non-uniform, the heat generated in the nanoparticle is non-uniform too. However, the temperature distribution within the nanoparticle is relatively uniform with differences less than 1°C, owing to the ultrafast thermal diffusion at the nanoscale.[43]

The steady state temperature of nanoparticles depends not only on the absorbed light power but also on the thermal diffusion efficiency of the surrounding medium. The power absorbed by a nanoparticle is given by $Q = \sigma_{abs} I$, where σ_{abs} and I are the absorption cross-section of the nanoparticle and the incident light intensity, respectively. The heat diffusion efficiency of the surrounding medium is characterized by its thermal conductivity. For nanoparticles dispersed in liquid solvents and under continuous light radiation, the steady temperature difference between nanoparticles and the surrounding environment is relatively small because liquids usually have relatively larger thermal conductivities and continuous light does not have very high intensities. For example, when a gold nanosphere sample with diameter of 20 nm is dispersed in water and illuminated at its resonance wavelength with an irradiance of $I = 1$ mWμm^{-2}, the temperature of the Au nanosphere is only ~5°C higher than that of water.[40] In contrast, if the surrounding medium of nanoparticles has a lower thermal conductivity, the steady temperature difference can be relatively large.[40] Moreover, when nanoparticles are illuminated with pulse lasers with very short pulse duration and high pulse rate, the temperature of nanoparticles can be much higher than that of the surrounding environment. Besides the local heating effect, the photothermal conversion of plasmonic nanoparticles can also heat reaction systems. For example, a 2 mL Au nanorod solution with an extinction value of 2 at the longitudinal plasmon peak and under resonant illumination at a laser light power of 2 W can increase the solution temperature from room temperature to above 70°C.[42]

Since the reaction rate constant of a chemical reaction is given by Arrhenius equation, $k = A\exp(-E_a/RT)$, and the photothermal conversion of plasmonic nanoparticles can increase the particle temperature and the reaction system temperature, photothermal conversion can therefore accelerate the reaction rate of chemical reactions. Moreover, the local heating effect, that is, the temperature of a metal nanoparticle being higher than that of its surrounding environment, can increase the selectivity of chemical reactions, because many reaction intermediates can quickly desorb and diffuse to the surrounding region at a lower temperature and will not react further.

4. Plasmon-assisted Organic Synthesis

Plasmonic catalysis was first observed in the degradation of the volatile organic compound HCHO under the illumination of visible light at ambient temperature.[32] Soon after, researchers realized that plasmonic catalysis is not limited to the degradation of organic pollutants. To date, plasmonic photocatalysts have been applied to various chemical reactions at moderate temperatures under UV- and visible-light illumination. Among them, the most important example is organic transformations, such as oxidation of alcohols, imine and amine synthesis and Miyaura–Suzuki reactions.

4.1 Selective oxidation of alcohols

Since aldehydes are very important organic intermediates or valuable components in the industries of perfumery, dyes, pharmaceuticals and agrochemicals,[44] selective oxidation of aromatic alcohols to aldehydes becomes a paramount process for both laboratory and commercial fine chemicals industries. However, the oxidation of alcohols is difficult to occur at ambient temperature. Moreover, traditional semiconductor photocatalysts, because of their low valence bands, give rise to a relatively low selectivity to aldehydes.[10] Plasmonic photocatalysts exhibit high activity and selectivity for the conversion of alcohols to aldehydes under visible-light illumination and therefore have been widely used in the selective oxidation of alcohols.[45–62]

Benzyl alcohol has often been used as a model oxidation reaction to test the catalytic activity and selectivity of plasmonic photocatalysts. The photocatalytic oxidation of benzylic alcohols in the presence of H_2O_2 on pure Au nanoparticles under a 530 nm light emitting diode (LED) illumination was demonstrated by Hallett-Tapley et al.[47] Pure colloidal Au nanoparticles were found to have a high photocatalytic activity under the illumination of 530 nm LEDs. Hot-electron transfer was used to explain the plasmon-enhanced photocatalytic activity. However, because the reaction was carried out on a drop of the nanoparticle solution in this study, the amount of the reactant molecules is very small and the catalyst is not convenient to be recycled. To overcome this disadvantage, the authors used Au nanoparticles supported on hydrotalcite and on metal oxides, including γ-Al_2O_3, n-Al_2O_3 and ZnO as photocatalysts in a later work (Eq (1)).[58] The overall efficiency of alcohol oxidation was found to be strongly dependent on the nature of the support. Hydrotalcite-supported Au nanoparticles give rise to the highest conversion of acetophenone and benzaldehyde from sec-phenethyl and benzyl alcohols, respectively.

$$\text{Ar-CH(OH)-R} + H_2O_2 \xrightarrow[\text{530 nm LED}]{\text{Au/support}} \text{Ar-C(O)-R} + 2H_2O \quad (1)$$

R = H, CH_3
support = hydrotalcite, γ-Al_2O_3, n-Al_2O_3, ZnO

Since the catalytic activity of Au nanoparticles with sizes larger than 2 nm, which is nearly the smallest size for nanoparticles that can support prominent plasmon resonance, is not high, bimetallic nanostructures have been prepared and used in plasmonic catalysis to improve the catalytic activity. Huang et al. synthesized Au/Pd nanowheels, for which each Au nanoplate core was surrounded by Pd, and successfully applied this catalyst to photocatalyze the oxidation of benzyl alcohol under Xe-lamp illumination.[54] Under the same experimental conditions, the introduction of light illumination clearly accelerates the reaction rate. For a 6h reaction, the conversion of benzyl alcohol is promoted from 18.4% under conventional heating to 97.7% under light illumination. This light-enhanced catalytic performance is attributed to the unique heterostructure and plasmon resonance of the Au/Pd nanowheels. Unfortunately, the authors did not study which effect, photothermal conversion or hot-electron injection, produces the enhancement.

For pure plasmonic metal and bimetallic nanostructures, light harvesting and catalysis are both realized by metals. Plasmonic metals can also be integrated with semiconductors. In (plasmonic metal)/semiconductor hybrid nanostructures, light can be harvested by plasmonic metals and/or semiconductors and catalytic reactions take place on both semiconductors and metals. TiO_2 is a n-type semiconductor and has been widely used as photocatalysts. The conduction-band of TiO_2 is only slightly higher than the Fermi level of Au. The hot electrons generated by plasmon in Au can therefore inject into the conduction-band of TiO_2 readily. Various types of TiO_2 samples have been employed as supports for plasmonic Au nanoparticles. Au nanoparticles smaller than 5 nm in diameter loaded on P25 can serve as plasmonic photocatalysts for the aerobic oxidation of benzylic alcohols in toluene under visible-light illumination ($\lambda > 450$ nm).[49] Compared to Au/anatase and Au/rutile counterparts, Au/P25 has the highest photocatalytic activity. The high catalytic activity obtained on Au/P25 is attributed to the consecutive electron transfer in the Au/rutile/anatase contact site (Fig. 5). The plasmon-induced hot electrons first inject from Au to the tightly bound rutile. The electrons then transfer from rutile to well-conjugated anatase and reduce O_2 to O_2^- on the surface of anatase. O_2^- abstracts the H atom from the OH group of benzylic alcohols and transfers the electron to the remaining part of the substrate. The negatively charged intermediate species is readily adsorbed on the electron-deficient Au nanoparticle and transfers an electron to the Au nanoparticle with the abstraction of a α-H atom from the methylene group ($-CH_2-$). More importantly, Au/P25 can give a relatively high catalytic efficiency for the aerobic oxidation of benzylic alcohols (Eq. (2)) directly under sunlight illumination. In addition to the reactions in organic solvents, Au/TiO_2 can also act as an efficient photocatalyst for the selective oxidation of benzylic alcohols with O_2 in water under the visible-light illumination of a 300 W Xe lamp (Eq. (3)).[45] When the cationic surfactant stearyltrimethylammonium chloride was added to the photocatalytic reaction system, the reaction rate was significantly enhanced by 3.3–5.7 folds. The photocatalytic activity is wavelength-dependent and displays a spectrum similar to the

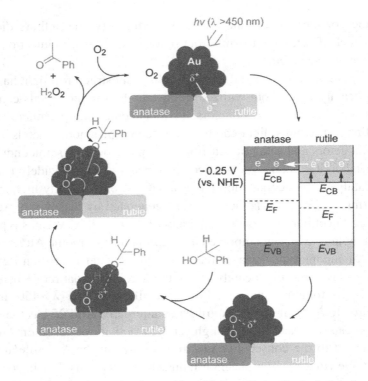

Fig. 5. Proposed mechanism for the aerobic oxidation of benzylic alcohol on the Au/P25 catalyst under visible-light illumination. Reproduced with permission.[49]

diffuse reflection spectrum of the Au nanoparticles, indicating that the enhanced photocatalytic activity stems from the plasmon. However, Au/ZrO_2 did not show any photocatalytic activity for the selective oxidation of benzylic alcohols under the same conditions. The different photocatalytic performances between Au/TiO_2 and Au/ZrO_2 are attributed to the fact that the conduction-band of ZrO_2 is much higher than the Fermi level of Au, which inhibits hot-electron transfer from Au nanoparticles to ZrO_2.

R^1 = H, CH_3, MeO, Cl
R^2 = H, CH_3, Ph

(2)

$$\text{PhCH(OH)R} \xrightarrow[\text{H}_2\text{O, STMAC, O}_2]{\text{Au/TiO}_2,\ >420\text{ nm, 300 W Xe lamp}} \text{PhC(O)R}$$

R = H, CH$_3$ STMAC = stearyltrimethylammonium chloride   (3)

To further increase the activity and decrease the cost of the catalysts, Au-based alloy nanoparticles supported on semiconductors have been utilized as photocatalysts. For example, Au nanoparticles alloyed with active transition metals like Cu, Pt, Ag and Pd supported on P25 were successfully synthesized. Au/Cu alloy nanoparticles supported on P25 can efficiently photocatalyze aerobic oxidation of benzylic alcohols under visible-light illumination.[53] The study showed that hot electrons generated by plasmon excitation did not inject into P25 but transferred to oxidized Cu atoms on the surface. P25 therefore only functioned as the support. The oxidation of alcohols took place on the Au/Cu alloy nanoparticles. The surface Cu atoms oxidized by O$_2$ during the reaction were successfully reduced by hot electrons generated by plasmon excitation. This maintained the alloying effect and promoted the aerobic oxidation of benzylic alcohols under visible-light illumination at ambient temperature. Moreover, Au nanoparticles alloyed with Pd were also shown to be an efficient plasmonic photocatalyst for the oxidation of benzyl alcohols when they were supported on ZrO$_2$,[56] which is different from Au/ZrO$_2$ that was inactive for the oxidation of benzyl alcohol under visible-light illumination.[45] This is because in the alloyed nanoparticles the plasmon-excited electrons can transfer to Pd to take part in the reaction.

CeO$_2$, a reducible metal oxide, can play the role of TiO$_2$ as the support for Au nanoparticles. Au/CeO$_2$ was prepared by photochemical deposition of H$_4$AuCl$_4$ on CeO$_2$ in the presence of citric acid as the reducing agent. The obtained Au/CeO$_2$ exhibited an absorption peak at 550 nm, which corresponds to the plasmon resonance of Au nanoparticles.[46] Under the illumination of green light from a 530 nm LED, Au/CeO$_2$ catalyzed the transformation of benzyl alcohol into benzaldehyde in the presence of O$_2$, with a selectivity higher than 99%. One should note that the reaction was carried out in water without the assistance of any surfactant. Other aromatic alcohols and their corresponding aldehydes also showed high conversion yields and selectivities (Eq. (4)). The external surface area of Au nanoparticles was found to play a crucial role in determining the activity of the photocatalyst. Au nanoparticles with average sizes larger than 30 nm were very active under the illumination of visible light. The size of Au nanoparticles loaded on CeO_2 can be increased by increasing the photo-deposition time.[50] Moreover, under the illumination of a 530-nm LED, the Au/CeO_2 can efficiently catalyze the conversion of 4-aminobenzyl alcohol to 4-aminobenzaldehyde despite the presence of an amine functional group (Eq. (5)). The plasmon resonance of Au nanoparticles were found to be responsible for the photocatalytic activity under visible-light illumination. The functioning mechanism of plasmon was attributed to the hot-electron effect. The hot electrons transferred from the Au nanoparticles to the conduction band of CeO_2 to take part in the reactions, and the resultant electron-deficient Au oxidized organic compounds to recover to its original metallic state.

$$R\text{-}C_6H_4\text{-}CH_2OH \xrightarrow[H_2O,\ 530\text{-nm LED},\ O_2]{Au/CeO_2} R\text{-}C_6H_4\text{-}CHO \qquad (4)$$
$$R = H,\ o\text{CH}_3,\ m\text{CH}_3,\ m\text{Cl},\ p\text{CH}_3,\ p\text{Cl}$$

$$NH_2\text{-}C_6H_4\text{-}CH_2OH \xrightarrow[H_2O,\ 530\text{-nm LED},\ O_2]{Au/CeO_2} NH_2\text{-}C_6H_4\text{-}CHO \qquad (5)$$

In addition to Au nanoparticles loaded on CeO_2 support, Au–CeO_2 core–shell nanostructures were also used as a photocatalyst for the oxidation of benzyl alcohol.[59] The Au@CeO_2 nanostructures were prepared in two steps. Au nanocrystals were first prepared by a wet-chemistry method, which allows for precise control of the size and shape of the nanocrystals and therefore their plasmon wavelength. CeO_2 shell was then grown on the pre-synthesized Au nanocrystals through heterogeneous nucleation and growth. Because the plasmon resonance of the core–shell nanostructures can be readily varied, sunlight in the entire visible region can be harvested for the catalytic reaction by mixing together the nanostructures with different plasmon wavelengths. The Au@CeO_2 nanostructures displayed a high photocatalytic activity for the selective aerobic oxidation of benzyl alcohol to benzaldehyde in acetonitrile under the visible-light illumination of a Xe lamp or a 671 nm laser. Both the photothermal and

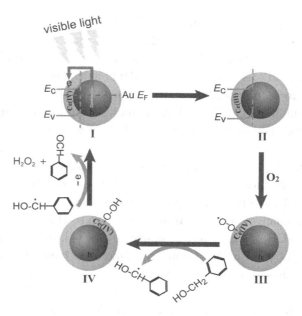

Fig. 6. Proposed mechanism for the selective aerobic oxidation of benzyl alcohol to benzaldehyde over the Au@CeO$_2$ nanostructure catalyst under visible-light illumination.

hot-electron effects were found to contribute to the photocatalytic activity. The contribution of the hot-electron effect was estimated to be ~53%. The proposed photocatalytic mechanism is shown in Fig. 6. The electrons injected into CeO$_2$ are trapped to form Ce(III) species. The enrichment of Ce(III) species facilitates the catalytic oxidation of benzyl alcohol into benzaldehyde with O$_2$.[63] Specifically, the adsorbed O$_2$ molecules on the surface of the CeO$_2$ shell can coordinate with the oxygen-deficient Ce(III), forming the Ce-coordinated superoxide species, Ce(IV)–O–O·. These superoxide species evolve into cerium hydroperoxide by abstraction of α-H from the methylene group (–CH$_2$–) of benzyl alcohol. Cerium hydroperoxide can then combine with a proton from the dehydrogenated species to produce H$_2$O$_2$, yielding benzaldehyde as the product. An electron is simultaneously transferred to the Au nanocrystal core to recover the charge balance.

Au nanoparticles supported on zeolite can also efficiently catalyze the oxidation of aromatic alcohols to aldehydes under visible-light illumination at ambient temperature.[48] Compared with thermally accelerated reactions, the conversion and the selectivity to aldehydes were clearly improved when the reactions were catalyzed by Au/zeolite under visible-light illumination. The enhancement was found to be dependent on the molecular polarity. The stronger the molecular polarity is, the larger the enhancement obtained by visible-light illumination. The enhancement was confirmed to arise from the plasmon-induced transient polarization of Au nanoparticles and the hot-electron effect. The proposed catalytic mechanism is shown in Fig. 7. Benzyl alcohol molecules are first adsorbed on the surface of the zeolite support through the formation of hydrogen bonds. Upon visible-light illumination, Au nanoparticles are transiently polarized owing to the plasmon resonance, resulting in a relatively high electronegativity. An α–H atom from the methylene group (–CH_2–) can then be abstracted by the polarized Au nanoparticle with the formation of a Au–H bond.[64] On the other hand, the hot electrons generated by plasmon decay can inject into O_2, forming activated O_2^- species. These species then remove H atoms from the Au–H bonds to yield water as the byproduct. At the same time, the electron-deficient Au nanoparticles are replenished by the electrons released from the adsorbed benzyl alcoholic compounds to form benzyl alcoholic radicals. Subsequently, these benzyl alcoholic radicals automatically release protonic hydrogen from the

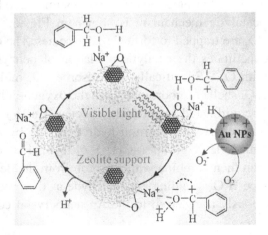

Fig. 7. Proposed mechanism for the selective aerobic oxidation of benzyl alcohol to benzaldehyde over the Au/zeolite photocatalyst under visible-light illumination. Reproduced with permission.[48]

hydroxyl groups (–OH) to form the C=O bond, followed by the desorption of the product benzaldehyde from the zeolite support.

In addition to aromatic alcohols, the selective conversion of 2-propanol into acetone has also been realized by plasmonic catalysis.[51,55,61] Au nanorods with varying aspect ratios supported on TiO_2 were utilized as the photocatalysts for the oxidation of 2-propanol under the illumination of visible-light.[55] Under broadband light illumination, the photocatalytic activity was enhanced by 2.8 times in terms of acetone production, which was mainly attributed to the plasmon resonance of Au nanorods. Unfortunately, the functions of plasmon and the catalytic mechanism are not investigated in this work. The photocatalytic activity of Au/TiO_2 for the oxidation of 2-propanol can be further improved by loading co-catalysts.[51] Au/TiO_2-M (M stands for a metal co-catalyst of Ir, Cu, Ag, Au, Rh, Ru, Pd or Pt) catalysts were prepared by the photo-deposition of a co-catalyst on TiO_2, followed by the loading of Au colloids.[51] The Au/TiO_2-M catalysts exhibited photocatalytic activities for the conversion of 2-propanol to H_2 and acetone (Eq. (6)) under the illumination of visible-light form a Xe lamp. The photocatalytic activity of the Au/TiO_2-M samples was found to be highly dependent on the nature of the co-catalyst. Au/TiO_2–Pt showed the highest photocatalytic activity, which was 5–9 times larger than that of Au/TiO_2. The apparent quantum efficiency (AQE) obtained with the Au/TiO_2–Pt catalyst was in good agreement with the plasmon resonance of the Au nanocrystals (Fig. 8(a)). The functioning mechanism of the plasmon resonance of the Au nanocrystals was believed to be the hot-electron effect. The hot electrons first inject from the Au nanocrystals into the conduction band of TiO_2 and then transfer to the co-catalyst to take part in the reaction (Fig. 8(b)).

$$H_3C-\underset{\underset{H}{|}}{\overset{\overset{OH}{|}}{C}}-CH_3 \xrightarrow[\text{400–600 nm, Xe lamp}]{\text{Au/TiO}_2\text{-M}} H_3C-\overset{\overset{O}{\|}}{C}-CH_3 + H_2$$

M = Ir, Cu, Ag, Au, Rh, Ru, Pd, Pt (6)

4.2 Imine and amine synthesis

Imines are important versatile intermediates for fine chemicals and pharmaceuticals. Some of them are under mass production, e.g. dyes.[65] Imines can be synthesized through the condensation of amines with aldehydes or

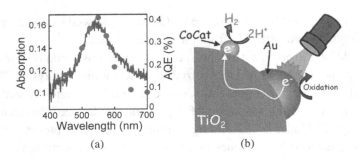

Fig. 8. (a) Absorption (black line) and AQE (red circles) spectra of the Au/TiO$_2$-Pt catalyst. (b) Proposed mechanism for the production of H$_2$ from 2-propanol in aqueous suspensions of Au/TiO$_2$-M under visible-light illumination. Reproduced with permission.[51]

direct oxidation of amines.[66] Some of these synthesis processes involve the use of stoichiometric amounts of harmful or explosive reagents as well as the emission of the corresponding amounts of pollutants.[67,68] Plasmonic catalysis provides a green and environment friendly method to the synthesis of imines under mild conditions.[56,69,70] Selective aerobic oxidation of amines to imines has been successfully catalyzed by Au nanoparticles loaded on various metal oxides, including TiO$_2$, WO$_3$, ZnO, In$_2$O$_3$ and SrTiO$_3$, under visible-light illumination at ambient temperature.[69] Although the overall conversions of amines are very low (from 0.1 to 4.5%) in a 24 h reaction (Table 1), very high selectivities (>99%) to imines are obtained on Au/rutile–TiO$_2$. The size of Au nanoparticles was found to play a determining role on the photocatalytic activity, because the increase in the Au nanoparticle size led to the increase in the plasmon resonance strength with a reduction in the Au nanoparticle surface area. The optimum diameter of the Au nanoparticles was found to be ~6.7 nm. The electrochemical study showed that the action spectrum of the initial rate of the electrode potential change $(dE/dt)_0$ shows a good resemblance to the absorption spectrum of Au/rutile–TiO$_2$ (Fig. 9(a)), indicating that the plasmon resonance is the driving force for amine oxidation. The proposed photocatalytic mechanism is shown in Fig. 9(b). The plasmon-excited electrons transfer from the Au nanoparticles to rutile TiO$_2$, which confers the oxidation ability on the Au nanoparticles. Owing to the mild oxidation ability, amines adsorbed on the Au nanoparticle surface are efficiently oxidized to imines without over-oxidation. On the other hand, O$_2$ is reduced by the

Table 1: Amine oxidation by Au/rutile–TiO$_2$[a].

Amine	Yield/mmol	Conversion %	Selectivity %	TON
X = H	1.06	4.5	>99	5.7×10^3
X = Me	0.43	2.1	>99	2.3×10^3
X = Cl	0.12	0.7	>99	6.5×10^2
X = OMe	0.49	2.7	>99	2.6×10^3
R = Me	0.03	0.1	>99	1.7×10^2
R = CH$_2$Ph	0.09	0.3	>99	4.7×10^2

Note: [a]Visible-light illumination (λ > 430 nm) for 24 h.

transferred electrons in the conduction-band of rutile TiO$_2$, which is believed to be the rate-determining step in the entire oxidation reaction. The selective oxidation of benzylamines to imines was also demonstrated on Au–Pd/ZrO$_2$ plasmonic photocatalysts under visible-light illumination (Eq. (7))[56]

(7)

Target imines can also be obtained by hydroamination of alkynes with anilines on a plasmonic photocatalyst, Au/TiO2–N, which represents Au nanoparticles supported on nitrogen-doped TiO$_2$, in toluene under visible-light illumination (Eq. (8)).[70] A high yield and selectivity were achieved at ambient temperature. The photocatalytic activity under the illumination of visible-light was attributed to the plasmonic effect of the Au nanoparticles. Under visible-light illumination, the plasmon-induced hot electrons inject from the Au nanoparticles into the conduction-band of TiO$_2$–N,

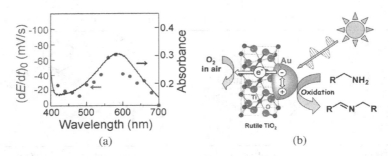

Fig. 9. (a) Absorption spectrum (black line) and action spectrum (violet circles) of the Au/rutile–TiO$_2$ catalyst. (b) Proposed mechanism for the aerobic oxidation of amines to imines on Au/rutile–TiO$_2$ under visible-light illumination. Reproduced with permission.[69]

producing positively charged Au nanoparticles. The nucleophilic aniline can interact with the positively charged Au nanoparticles to form N-centered radical cations, which leads to its activation for the electrophilic attack on the electron-rich sites of alkynes. At the same time, alkynes can be activated via the interaction of the terminal C–H bond of alkynes with the active Ti^{3+} sites formed owing to N-doping in the support. The next step most probably entails the addition of the two activated reactants: the N atom links to the C atom of the phenyl ring, while a H atom is added onto the terminal C atom of the C≡C bond. The final step is the desorption of the product molecules from the support surface. In addition, one-pot synthesis of propargyl amines through the ternary coupling of aldehyde, alkyne and secondary amine has been realized using a Au/ZnO catalyst under the illumination of a 530 nm LED (Eq. (9)). The Au/ZnO catalyst gives rise to a rapid and selective formation of the products with good yields under ambient conditions.[71]

$$R^1\text{—}{\equiv}\text{—}H + H_2N\text{—}R^2 \xrightarrow[\text{halogen lamp}]{\text{Au/TiO}_2\text{-N}} \underset{R^1}{\overset{N(R^2)}{\underset{|}{C}}}\text{—}CH_3$$

R^1 = Ph-(CH$_2$)$_2$, Ph-CH$_2$, Ph, CH$_3$(CH$_2$)$_5$, Br-Ph, CH$_3$O-Ph, CH$_3$-Ph, Ph-C≡C-CH$_3$, Ph-C≡C-Ph

R^2 = Ph, Cl-Ph, Br-Ph, CH$_3$-Ph, CH$_3$O-Ph, C$_6$H$_{11}$, Ph-NH-CH$_3$

(8)

$$R^1 = CH_2, C_2H_4, OCH_2$$
$$R^2 = H, C_6H_5, CH(CH_3)_2$$

(9)

4.3 Miyaura–Suzuki coupling reactions

Miyaura–Suzuki coupling reactions play a paramount role in organic synthesis. It can produce biaryls straightforwardly and conveniently. Great efforts have been made to use transition metals to catalyze Miyaura–Suzuki coupling reactions, especially homogeneous palladium catalysts. Very recently, the interest in palladium catalysts has shifted to plasmonic bimetallic Au/Pd nanostructures, which can catalyze Miyaura–Suzuki coupling reactions under light illumination at ambient temperature and therefore is very promising for practical and industrial applications.[54,56,72–77] Bimetallic Au/Pd nanostructures have been used as photocatalysts for Suzuki coupling reactions (Eq. (10)) under laser and solar illumination.[72] The Au/Pd nanostructures were prepared by the reduction of Pd in the presence of as-grown Au nanorods. The plasmon resonance of the Au nanorods can effectively harvest light energy, while the Pd nanoparticles are the catalytic active centers for Suzuki coupling reactions. Since the plasmon resonance of Au nanorods can be readily tailored from the visible to NIR region, the mixture of Au/Pd nanostructures with different plasmon wavelengths can absorb almost all visible and NIR solar energy. For various substituted reactants, very high yields were obtained in 2 h reactions under solar illumination at ambient temperature (Table 2). Smaller Au/Pd nanostructures were found to show a higher photocatalytic performance compared with larger Au/Pd nanostructures, because the non-radiative channel dominates the decay of the plasmon in the smaller nanostructures. Systematic mechanism studies showed that both the photothermal and hot-electron effects contribute to the high photocatalytic activity, and that the hot-electron effect plays a dominant role. In addition,

Table 2: Yields of Suzuki coupling reactions in the presence of the mixture of Au/Pd nanostructures under solar illumination[a]

Entry	Halide	Boronic acid	Yield [%]
1	Ph–I	Ph–B(OH)$_2$	97
2	Ph–Br	Ph–B(OH)$_2$	99
3	Ph–Br	2-CH$_3$-C$_6$H$_4$–B(OH)$_2$	99
4	Ph–Br	3-CH$_3$-C$_6$H$_4$–B(OH)$_2$	96
5	Ph–Br	4-CH$_3$-C$_6$H$_4$–B(OH)$_2$	99
6	2-OCH$_3$-C$_6$H$_4$–Br	2-CH$_3$-C$_6$H$_4$–B(OH)$_2$	50
7	2-OCH$_3$-C$_6$H$_4$–Br	3-CH$_3$-C$_6$H$_4$–B(OH)$_2$	67
8	2-OCH$_3$-C$_6$H$_4$–Br	4-CH$_3$-C$_6$H$_4$–B(OH)$_2$	70
9	3-OCH$_3$-C$_6$H$_4$–Br	2-CH$_3$-C$_6$H$_4$–B(OH)$_2$	99
10	3-OCH$_3$-C$_6$H$_4$–Br	3-CH$_3$-C$_6$H$_4$–B(OH)$_2$	99
11	3-OCH$_3$-C$_6$H$_4$–Br	4-CH$_3$-C$_6$H$_4$–B(OH)$_2$	99
12	4-CH$_3$O-C$_6$H$_4$–Br	2-CH$_3$-C$_6$H$_4$–B(OH)$_2$	99
13	4-CH$_3$O-C$_6$H$_4$–Br	3-CH$_3$-C$_6$H$_4$–B(OH)$_2$	76
14	4-CH$_3$O-C$_6$H$_4$–Br	4-CH$_3$-C$_6$H$_4$–B(OH)$_2$	55

[a]Au/Pd nanostructure mixture solution, 1 mL; NaOH, 10 mg; cetyltrimethylammonium bromide, 36 mg; H$_2$O, 1 mL; reactant amounts of entries 1–5, 0.08 mmol halide and 0.08 mmol boronic acid; reactant amounts of entries 6–14, 0.04 mmol halide and 0.04 mmol boronic acid; solar power, 565 W/m^{-2}; environment temperature, 30°C; reaction time, 2 h.

bimetallic Au/Pd nanowheels have also been employed in photocatalytic Miyaura–Suzuki coupling reactions under light illumination. They exhibit good catalytic performances.[54]

$$\underset{R^1}{\bigcirc}-X + (HO)_2B-\underset{}{\bigcirc}-R^2 \xrightarrow[\text{or sunlight}]{\text{Au/Pd nanorods}} \underset{R^1}{\bigcirc}-\underset{}{\bigcirc}-R^2$$

X = I; R^1 = H; R^2 = H

X = Br; R^1 = 2-OCH$_3$, 3-OCH$_3$, 4-OCH$_3$; R^2 = 2-CH$_3$, 3-CH$_3$, 4-CH$_3$ (10)

Alloyed Au/Pd nanoparticles supported on ZrO_2 have also been found to catalyze Miyaura–Suzuki coupling reactions under visible-light illumination (Eq. (11)) to produce the target products in high yields.[56,74] The contributions of light and thermal effects to the conversion efficiencies were studied carefully by varying the light intensity and wavelength. As shown in Fig. 10(a), the linear relationship between the conversion yield and the light intensity clearly demonstrated that the enhancement results from the light illumination. Moreover, the contribution from light was found to be dependent on the light wavelength (Fig. 10(b)). The light, spectrally covering the plasmon wavelength of the Au/Pd nanoparticles gives rise to more prominent enhancements compared with the light that is spectrally away from the plasmon wavelength. These results indicated clearly that the enhancement of the photocatalytic activity stems mainly from the plasmon resonance of the Au/Pd nanoparticles.

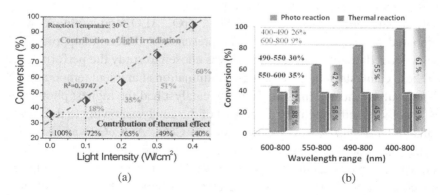

Fig. 10. Dependence of the photocatalytic activity of the Au/Pd alloy nanoparticles for Miyaura–Suzuki coupling reaction on the light intensity (a) and wavelength (b). Both the light-driven reaction and the reaction in the dark were conducted at 30 ± 1°C. Reproduced with permission.[56]

$$\text{R}^1\text{-}\!\!\!\!\!\bigcirc\!\!\!\!\!\text{-I} + (\text{HO})_2\text{B-}\!\!\!\!\!\bigcirc\!\!\!\!\!\text{-R}^2 \xrightarrow[\text{(λ = 400–700 nm)}]{\text{Au/Pd/ZrO}_2 \text{ halogen lamp}} \text{R}^1\text{-}\!\!\!\!\!\bigcirc\!\!\!\!\!\text{-}\!\!\!\!\!\bigcirc\!\!\!\!\!\text{-R}^2$$

R^1 = 4-H, 2-CH$_3$, 3-CH$_3$, 4-CH$_3$, 4-OCH$_3$
R^2 = 4-H, 4-OCH$_3$, 4-CHO, 4-N(CH$_3$)$_2$ (11)

4.4 Other organic transformation reactions

In addition to the three types of reactions described above, plasmonic photocatalysts have also been utilized in a wide range of other reactions,[33,34,78–92] such as selective reduction reactions, hydroxylation of benzene to phenol, hydrogenation of styrene to ethylbenzene, oxidation of 9-anthraldehyde to anthraquinone, amide synthesis, selective oxidation of thiols to disulfides and conversion of aldehydes to esters. Zhu *et al.* have applied plasmonic catalysis in selective reduction reactions,[78,83,86] including reduction of nitro-aromatics to azo compounds, hydrogenation of azobenzene to hydroazobenzene, reduction of ketones to alcohols, and deoxygenation of epoxides to alkenes. These reduction reactions are important in organic synthesis and biological chemistry.[93–97] For example, deoxygenation of epoxides is an important step in the vitamin K cycle.[93,94] The reduction of ketones to alcohols is a significant transformation in organic synthesis and the chemical industry.[95,96] Azo compounds are widely used as dyes in the polymer, textile and food industries as well as in the production of medicines to treat inflammation and a type of arthritis.[97] Plasmonic photocatalysts composed of Au nanoparticles grown on various supports, including CeO$_2$, TiO$_2$, ZrO$_2$, Al$_2$O$_3$ and zeolite Y, were prepared by reducing HAuCl$_4$ in the presence of different supports. The selective reduction of styrene oxide was chosen to study the performances of these catalysts under visible-light illumination at ambient temperature.[86] The results indicated that the Au/CeO$_2$ catalyst had the highest photocatalytic activity and selectivity among all of the catalysts, owing to the strong linkage between the Au nanoparticles and CeO$_2$. Au/CeO$_2$ was also used as the photocatalyst in the other three types of selective reduction reactions shown in Table 3. Light illumination was found to prominently enhance the conversion yields of different reactant molecules. The enhancement is ascribed to the plasmon-induced hot electrons of the Au nanoparticles. The proposed catalytic mechanism is shown in Fig. 11. In the reaction, isopropanol

Table 3: Performance of Au/CeO$_2$ for the four photocatalytic reactions

Entry	Reagent	Products	Time	Temperature	Visible-light Conversion	Visible-light Selectivity
1	PhNO$_2$	Ph-N$^+$(O$^-$)=N-Ph	2 h	30°C	43.5%	96%
2	Ph-N=N-Ph	Ph-NH-NH-Ph	6 h	30°C	40%	78%
3	acetophenone	1-phenylethanol	24 h	30°C	31%	>99%
4	styrene oxide	styrene (PhCH=CH$_2$)	16 h	25°C	20%	88%

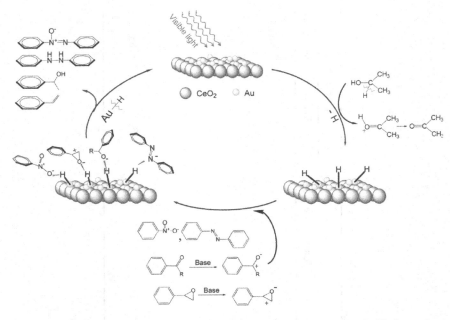

Fig. 11. Proposed catalytic mechanism in the selective reduction with the Au/CeO$_2$ catalyst. Reproduced with permission.[86]

is used as a solvent, which is a hydrogen donor. Under visible-light illumination, the Au nanoparticles are able to abstract hydrogen atoms from isopropanol, giving rise to transient Au–H species. With the assistance of the hot electrons generated by the plasmon resonance, the Au–H species can break the double bonds (N=O, C=O, N=N) or the epoxide bond leading to the hydrogenation or deoxygenation, where the hydrogen atom of the Au–H species is consumed and the final reduction products are released. The enhanced local electromagnetic field is also believed to have a promotion effect on the activation of the double bonds or the epoxide bond. Moreover, the photocatalytic activity of Au/CeO$_2$ was found to be dependent on the wavelength of light.[83,86] Shorter wavelength light gives rise to a higher photo-reducing capability. This dependence is ascribed to the fact that the energy of the hot electrons relies on the illumination light wavelength. Only the electrons with energies higher than the reduction potentials of the reactants are able to reduce them (Fig. 12).

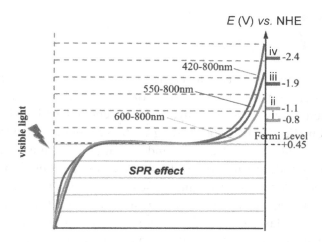

Fig. 12. Schematic of the band structure of Au nanoparticles with different illumination wavelengths and reduction potentials of typical reactants. (i) nitrobenzene; (ii) azobenzene; (iii) acetophenone; (iv) styrene oxide. Reproduced with permission.[86]

The one-step hydroxylation of benzene to phenol is challenging, because this reaction usually has a low yield and selectivity. The application of plasmonic photocatalysts in the hydroxylation reaction of benzene can provide a valuable solution to overcome this problem. Huang *et al.* have prepared plasmonic noble metal photocatalysts M/TiO$_2$ (M = Pt, Ag and Au) by a facile *in situ* method and utilized them to catalyze the hydroxylation of benzene (eq. 12) under visible-light illumination ($\lambda > 400$ nm).[82] Among the three types of photocatalysts, Au/TiO$_2$ exhibited the best photocatalytic performance. The content of Au loading was found to play a key role in determining the photocatalytic activity and selectivity. 2 wt.% Au loading showed the best dispersivity and photocatalytic performance with 63% yield and 91% selectivity. The photocatalytic activity under visible-light illumination was attributed to the plasmon-induced hot-electron effect. The hot electrons transfer from the Au nanoparticles to the TiO$_2$ particle, and the electron-deficient Au nanoparticles oxidize phenoxy anions to form phenoxy radicals that further oxidize benzene to phenol. Moreover, the photocatalytic hydroxylation of benzene in aqueous solutions under sunlight with Au/TiO$_2$ can be considerably improved when the reaction atmosphere was CO$_2$.[81]

$$\text{C}_6\text{H}_6 \xrightarrow[\text{Xe lamp } (\lambda > 400 \text{ nm})]{\text{M/TiO}_2 \text{ (M = Au, Pt and Ag)}} \text{C}_6\text{H}_5\text{-OH} \qquad (12)$$

Ke *et al.* have prepared Au/zeolite photocatalysts through the reduction of $HAuCl_4$ in the presence of zeolite and used them to catalyze acetalization between benzaldehyde and 1-pentanol under visible-light illumination (Eq. (13)).[33] The loading of Au can obviously enhance the conversion yields of the reactants under visible-light illumination. The enhancement under visible-light illumination was attributed to the plasmon-enhanced local electromagnetic field, which can activate the polar C=O bond in benzaldehyde. Moreover, the redox reactions can also be catalyzed by plasmonic catalysts under visible-light illumination. Kominami *et al.* have studied the photocatalytic performance of Au/TiO_2–Ag for the conversion of nitrobenzene and 2-propanol to aniline and acetone under visible-light illumination.[85] In this reaction, nitrobenzene is reduced into aniline and 2-propanol is simultaneously oxidized into acetone (Eq. (14)). Similarly, under visible-light illumination, simultaneous conversion of chlorobenzene into benzene and 2-propanol into acetone was realized on a catalyst of Au@Pd nanostructures supported on TiO_2 (Eq. (15)).[88] In addition, Linic *et al.* have shown that plasmonic photocatalysts can be used in gas reactions at high temperatures.[34,89] On the Ag nanoparticle catalyst, visible-light illumination was found to increase the reaction rate of ethylene epoxidation by 4-fold, although the reaction temperature was kept the same.[34] A linear dependence of the photo-induced reaction rate on the illumination light intensity signifies that the reactions are driven by the hot electrons generated by plasmon decay. The selective epoxidation of propylene to propylene oxide on Cu@Cu_2O nanoparticles has also been demonstrated.[89] The selectivity to propylene oxide with light illumination was found to be clearly larger than that without light illumination. The increased selectivity was attributed to the reduction of the Cu_2O shell into Cu under light illumination, which was revealed by the extinction spectral change of the core–shell nanoparticle catalyst upon light illumination. The light-induced reduction of Cu_2O was caused by the hot electrons derived from the plasmon resonance of the Cu core.

$$\text{PhCHO} + \text{CH}_2=\text{CHCH}_2\text{OH} \xrightarrow[\text{halogen lamp} \atop (\lambda = 420\text{–}800 \text{ nm})]{\text{Au/zeolite}} \text{Ph-CH(OH)-O-CH}_2\text{CH}=\text{CH}_2 \quad (13)$$

$$\text{PhNO}_2 + 3\,\text{(CH}_3)_2\text{CHOH} \xrightarrow[\text{Xe lamp} \atop (\lambda = 450\text{–}600 \text{ nm})]{\text{Au/TiO}_2\text{-Ag}} \text{PhNH}_2 + 3\,\text{(CH}_3)_2\text{C=O} + 2\text{H}_2\text{O} \quad (14)$$

$$\text{PhCl} + \text{(CH}_3)_2\text{CHOH} \xrightarrow[\text{Xe lamp} \atop (\lambda = 450\text{–}600 \text{ nm})]{\text{Au@Pd/TiO}_2} \text{PhH} + \text{(CH}_3)_2\text{C=O} + \text{HCl} \quad (15)$$

5. Conclusions and Outlook

In summary, the introduction of LSPRs into photocatalysis can efficiently improve the photocatalytic activity and modify the product selectivity under visible-light illumination. Using plasmonic photocatalysts to achieve organic synthesis under visible-light illumination is promising, and the obtained results are encouraging. The understanding of the traditional photocatalysis has been recently updated by the facts that pure plasmonic metal nanoparticles without the presence of semiconductor components can have photocatalytic ability and that the plasmon resonances of noble metal nanoparticles can significantly improve the photocatalytic ability of adjacent semiconductors. However, the use of plasmonic photocatalysts in the synthesis of useful organic compounds is still a new field, and the types of reactions catalyzed by plasmonic photocatalysts are still limited. From our perspective, several challenges need to be overcome in the future to fully realize the potential of plasmonic catalysis.

First, the physical understanding of plasmonic catalysis requires more studies. Since hot electrons pay a pivotal role in plasmon-assisted photocatalysis, the relationship between the maximum energy of hot electrons and the distribution of hot electrons in energy have not been well understood. Moreover, the spatial distribution of hot electrons in plasmonic nanoparticles has also remained unknown. However, the energy and spatial distributions of hot electrons are of paramount importance to photocatalysis, because the energy distribution determines whether hot electrons have sufficient energy to drive the reaction, and the spatial distribution of hot electrons determines the active site in catalysis. In addition, the role played by plasmon-enhanced local electromagnetic field in reactions needs more efforts to understand.

Second, the dependence of the photocatalytic activity on the plasmon wavelength and the size and shape of plasmonic nanoparticles have not been fully understood. Although the dependence of photocatalytic activities on the illumination light wavelength has been investigated previously,[83,86] in these studies metal nanoparticles with fixed plasmon wavelengths were employed. As a result, the photocatalytic activity was found to be high when the illumination light wavelength matched with the plasmon wavelength, but the dependence of the photocatalytic activity on the plasmon wavelength was not unraveled. Understanding the latter case is important, because the energies of hot electrons are expected to be dependent on the plasmon wavelength. On the other hand, for nanoparticles with the same plasmon wavelength, we know that differently shaped nanoparticles generate different local electromagnetic field enhancements. Whether the amplitude and spatial distribution of the local field enhancement affect the photocatalytic activity warrants further investigation.

Third, so far, the contributions from the hot-electron and photothermal effects have been obtained quantitatively by comparing the reaction without light illumination at the same temperature as that of the same reaction under light illumination. However, plasmonic photothermal conversion is a local effect, which can make the temperature of nanoparticles much higher than that of the reaction system, especially when nanoparticles are surrounded by materials with low thermal conductivities, for example in catalytic gas reactions. Since reactions usually take place on the surface of nanoparticles and reaction rate constants are exponential functions of temperature, reactions can be greatly enhanced by the local plasmonic heating effect. How to measure the local temperature on the surface of metal nanocrystals has remained a great challenge. Experimental means are highly desired for accurately and reliably measuring the local temperature on the surface of metal nanocrystals. On the basis of the local temperature measurements, approaches and experiments can be designed to quantitatively separate the contributions of the hot-electron and photothermal effects to the catalytic activity.

Fourth, although the use of plasmonic catalysis has been applied to many reactions, the reaction types have still remained limited. There is a need to extend plasmonic catalysis to various practical reactions, for which more efforts are required to understand and control the photocatalytic performances in various reaction systems.

We believe that with more attention paid to this field, the challenges mentioned above can be overcome in the forthcoming years. The attack of these challenges will lead to the widespread applications of plasmonic photocatalysts in organic synthesis, which has far-reaching significance to the development of green, safe and low-cost chemical industries.

References

1. Y. Q. Qu and X. F. Duan. *Chem. Soc. Rev.* **42**, 2013, 2568–2580.
2. J. C. Colmenares and R. Luque, *Chem. Soc. Rev.* **43**, 2014, 765–778.
3. A. Shah, P. Torres, R. Tscharner, N. Wyrsch and H. Keppner, *Science* **285**, 1999, 692–698.
4. N. S. Lewis. *MRS Bull.* **32**, 2007, 808–820.
5. C. K. Prier, D. A. Rankic and D. W. C. MacMillan, *Chem. Rev.* **113**, 2013, 5322–5363.
6. W. G. Tu, Y. Zhou and Z. G. Zou. *Adv. Mater.* **26**, 2014, 4607–4626.
7. J. R. Ran, J. Zhang, J. G. Yu, M. Jaroniec and S. Z. Qiao, *Chem. Soc. Rev.* **43**, 2014, 7787–7812.
8. G. Liu, L. Z. Wang, H. G. Yang, H.-M. Cheng and G. Q. Lu, *J. Mater. Chem.* **20**, 2010, 831–843.
9. N. Zhang, S. Q. Liu and Y.-J. Xu. *Nanoscale* **4**, 2012, 2227–2238.
10. M. D. Xiao, R. B. Jiang, F. Wang, C. H. Fang, J. F. Wang and J. C. Yu, *J. Mater. Chem. A* **1**, 2013, 5790–5805.
11. M. J. Kale, T. Avanesian and P. Christopher, *ACS Catal.* **4**, 2014, 116–128.
12. R. B. Jiang, B. X. Li, C. H. Fang and J. F. Wang, *Adv. Mater.* **26**, 2014, 5274–5309.
13. C. L. Wang and D. Astruc, *Chem. Soc. Rev.* **43**, 2014, 7188–7216.
14. Q. Xiao, E. Jaatinen and H. Y. Zhu, *Chem. Asian J.* **9**, 2014, 3046–3064.

15. H. F. Cheng, K. Fuku, Y. Kuwahara, K. Mori and H. Yamashita, *J. Mater. Chem. A* **3**, 2015, 5244–5258.
16. S. C. Warren and E. Thimsen, *Energ. Environ. Sci.* **5**, 2012, 5133–5146.
17. S. Mubeen, J. Lee, N. Singh, S. Krämer, G. D. Stucky, M. Moskovits, *Nat. Nanotechnol.* **8**, 2013, 247–250.
18. Z. K. Zheng, T. Tachikawa and T. Majima, *J. Am. Chem. Soc.* **136**, 2014, 6870–6873.
19. W. B. Hou, W. H. Hung, P. Pavaskar, A. Goeppert, M. Aykol and S. B. Cronin, *ACS Catal.* **1**, 2011, 929–936.
20. G. V. Hartland, *Chem. Rev.* **111**, 2011, 3858–3887.
21. H. J. Chen, L. Shao, Q. Li and J. F. Wang, *Chem. Soc. Rev.* **42**, 2013, 2679–2724.
22. H. C. Van de Hulst, *Light Scattering by Small Particles* (Dover: New York, 1981).
23. C. Bohren, D. Huffman, *Absorption and Scattering of Light by Small Particles* (Wiley: New York, 1983).
24. A. Boltasseva and H. A. Atwater, *Science* **331**, 2011, 290–291.
25. T. Ming, H. J. Chen, R. B. Jiang, Q. Li and J. F. Wang, *J. Phys. Chem. Lett.* **3**, 2012, 191–202.
26. V. Giannini, A. I. Fernández-Domínguez, S. C. Heck and S. A. Maier, *Chem. Rev.* **111**, 2011, 3888–3912.
27. N. J. Halas, S. Lal, W.-S. Chang, S. Link and P. Nordlander, *Chem. Rev.* **111**, 2011, 3913–3961.
28. C. Sönnichsen, T. Franzl, T. Wilk, G. von Plessen and J. Feldmann, *Phys. Rev. Lett.* **88**, 2002, 077402.
29. C. Bauer, J.-P. Abid, D. Fermin and H. H. Girault, *J. Chem. Phys.* **120**, 2004, 9302–9315.
30. O. Ekici, R. K. Harrison, N. J. Durr, D. S. Eversole, M. Lee and A. Ben-Yakar, *J. Phys. D: Appl. Phys.* **41**, 2008, 185501.
31. D. Werner, A. Furube, T. Okamoto and S. Hashimoto, *J. Phys. Chem. C* **115**, 2011, 8503–8512.
32. X. Chen, H.-Y. Zhu, J.-C. Zhao, Z.-F. Zheng and X.-P. Gao, *Angew. Chem. Int. Edit.* **47**, 2008, 5353–5356.
33. X. G. Zhang, X. B. Ke, A. J. Du and H. Y. Zhu, *Sci. Rep.* **4**, 2014, 3805.
34. P. Christopher, H. L. Xin and S. Linic, *Nat. Chem.* **3**, 2011, 467–472.
35. C. D. Lindstrom and X.-Y. Zhu, *Chem. Rev.* **106**, 2006, 4281–4300.
36. J. D. Beckerle, M. P. Casassa, R. R. Cavanagh and E. J. Heilweil, J. C. Stephenson, *Phys. Rev. Lett.* **64**, 1990, 2090–2093.

37. P. L. Redmond, X. M. Wu and L. Brus, *J. Phys. Chem. C* **111**, 2007, 8942–8947.
38. X. M. Wu, P. L. Redmond, H. T. Liu, Y. H. Chen, M. Steigerwald and L. Brus, *J. Am. Chem. Soc.* **130**, 2008, 9500–9506.
39. L. Brus, *Acc. Chem. Res.* **41**, 2008, 1742–1749.
40. G. Baffou and R. Quidant, *Laser Photonics Rev.* **7**, 2013, 171–187.
41. J. J. Qiu and W. D. Wei, *J. Phys. Chem. C* **118**, 2014, 20735–20749.
42. H. J. Chen, L. Shao, T. Ming, Z. H. Sun, C. M. Zhao, B. C. Yang and J. F. Wang, *Small* **6**, 2010, 2272–2280.
43. G. Baffou and R. Quidant, C. Girard, *Appl. Phys. Lett.* **94**, 2009, 153109.
44. R. A. Sheldon and J. K. Kochi, *Metal-catalyzed oxidations of organic compounds* (Academic press: New York, 1981).
45. S.-i. Naya, A. Inoue and H. Tada, *J. Am. Chem. Soc.* **132**, 2010, 6292–6293.
46. A. Tanaka, K. Hashimono and H. Komonami, *Chem. Commun.* **47**, 2011, 10446–10448.
47. G. L. Hallett-Tapley, M. J. Silvero, M. González-Béjar, M. Grenier, J. C. Netto-Ferrerira and J. C. Scaiano, *J. Phys. Chem. C* **115**, 2011, 10784–10790.
48. X. G. Zhang, X. B. Ke and H. Y. Zhu, *Chem. Eur. J.* **18**, 2012, 8048–8056.
49. D. Tsukamoto, Y. Shiraishi, Y. Sugano, S. Ichikawa, S. Tanaka and T. Hirai, *J. Am. Chem. Soc.* **134**, 2012, 6309–6315.
50. A. Tanaka, K. Hashimoto and H. Kominami, *J. Am. Chem. Soc.* **134**, 2012, 14526–14533.
51. A. Tanaka, S. Sakaguchi, K. Hashimoto and H. Kominami, *ACS Catal.* **3**, 2013, 79–85.
52. Y. Shiraishi, H. Sakamoto, Y. Sugano, S. Ichikawa and T. Hirai, *ACS Nano* **7**, 2013, 9287–9297.
53. Y. Sugano, Y. Shiraishi, D. Tsukamoto, S. Ichikawa, S. Tanaka and T. Hirai, *Angew. Chem. Int. Edit.* **52**, 2013, 5295–5299.
54. X. Q. Huang, Y. J. Li, Y. Chen, H. L. Zhou, X. F. Duan and Y. Huang, *Angew. Chem. Int. Ed.* **52**, 2013, 6063–6067.
55. L. Q. Liu, S. X. Ouyang and J. H. Ye, *Angew. Chem. Int. Edit.* **52**, 2013, 6689–6693.
56. S. Sarina, H. Y. Zhu, E. Jaatinen, Q. Xiao, H. W. Liu, J. F. Jia, C. Chen and J. Zhao, *J. Am. Chem. Soc.* **135**, 2013, 5793–5801.
57. R. Sellappan, M. G. Nielsen, F. Gonzáez-Posada, P. C. K. Vesborg, I. Chorkendorff and D. Chakarov, *J. Catal.* **307**, 2013, 214–221.

58. G. L. Hallett-Tapley, M. J. Silvero, C. J. Bueno-Alejo, M. González-Béjar, C. D. McTiernan, M. Grenier, J. C. Netto-Ferreira and J. C. Scaiano, *J. Phys. Chem. C* **117**, 2013, 12279–12288.
59. B. X. Li, T. Gu, J. X. Wang, P. Wang, J. F. Wang and J. C. Yu, *ACS Nano* **8**, 2014, 8152–8162.
60. S. Sarina, S. Bai, Y. M. Huang, C. Chen, J. F. Jia, E. Jaatinen, G. A. Ayoko, Z. Bao and H. Y. Zhu, *Green Chem.* **16**, 2014, 331–341.
61. Y. Horiguchi, T. Kanda, K. Torigoe, H. Sakai and M. Abe, *Langmuir* **30**, 2014, 922–928.
62. T. T. Jiang, C. C. Jia, L. C. Zhang, S. R. He, Y. H. Sang, H. D. Li, Y. Q. Li, X. H. Xu and H. Liu, *Nanoscale* **7**, 2015, 209–217.
63. A. Abad, P. Concepción, A. Corma and H. García, *Angew. Chem. Int. Edit.* **44**, 2005, 4066–4069.
64. M. Conte, H. Miyamura, S. Kobayashi and V. Chechik, *J. Am. Chem. Soc.* **131**, 2009, 7189–7196.
65. P. G. Cozzi, *Chem. Soc. Rev.* **33**, **2004**, 410–421.
66. S.-I. Murahashi, *Angew. Chem. Int. Edit.* **34**, 1995, 2443–2465.
67. T. Mukaiyama, A. Kawana and Y. Fukuda, J.-i. Matsuo, *Chem. Lett.* 2001, 390–391.
68. K. C. Nicolaou, C. J. N. Mathison and T. Montagnon, *Angew. Chem. Int. Edit.* **42**, **2003**, 4077–4082.
69. S.-i. Naya, K. Kimura and H. Tada, *ACS Catal.* **3**, 2013, 10–13.
70. J. Zhao, Z. F. Zheng, S. Bottle, A. Chou, S. Sarina and H. Y. Zhu, *Chem. Commun.* **49**, 2013, 2676–2678.
71. M. González-Béjar, K. Peters, G. L. Hallett-Tapley, M. Grenier and J. C. Scaiano, *Chem. Commun.* **49**, 2013, 1732–1734.
72. F. Wang, C. H. Li, H. J. Chen, R. B. Jiang, L.-D. Sun, Q. Li, J. F. Wang, J. C. Yu and C.-H. Yan, *J. Am. Chem. Soc.* **135**, 2013, 5588–5601.
73. Y. X. Hu, Y. Z. Liu, Z. Li and Y. G. Sun, *Adv. Funct. Mater.* **24**, 2014, 2828–2836.
74. Q. Xiao, S. Sarina, E. Jaatinen, J. F. Jia, D. P. Arnold, H. W. Liu and H. Y. Zhu, *Green Chem.* **16**, 2014, 4272–4285.
75. S. T. Gao, N. Z. Shang, C. Feng, C. Wang and Z. Wang, *RSC Adv.* **4**, 2014, 39242–39247.
76. R. Long, Z. L. Rao, K. K. Mao, Y. Li, C. Zhang, Q. L. Liu, C. M. Wang, Z.-Y. Li, X. J. Wu and Y. J. Xiong, *Angew. Chem. Int. Edit.* **54**, 2015, 2425–2430.
77. M. C. Wen, S. Takakura, K. Fuku and K. Mori, H. Yamashita, *Catal. Today* **242**, 2015, 381–385.

78. H. Y. Zhu, X. B. Ke, X. H. Yang, S. Sarina and H. W. Liu, *Angew. Chem. Int. Edit.* **49**, 2010, 9657–9661.
79. S.-i. Naya, M. Teranishi, T. Isobe and H. Tada, *Chem. Commun.* **46**, 2010, 815–817.
80. C. J. B. Alejo, C. Fasciani, M. Grenier, J. C. Netto-Ferreira and J. C. Scaiano, *Catal. Sci. Technol.* **1**, 2011, 1506–1511.
81. Y. Ide, N. Nakamura, H. Hattori, R. Ogino, M. Ogawa, M. Sadakane and T. Sano, *Chem. Commun.* **47**, 2011, 11531–11533.
82. Z. K. Zheng, B. B. Huang, X. Y. Qin, X. Y. Zhang, Y. Dai and M.-H. Whangbo, *J. Mater. Chem.* **21**, 2011, 9079–9087.
83. X. B. Ke, S. Sarina, J. Zhao, X. G. Zhang, J. Chang and H. Y. Zhu, *Chem. Commun.* **48**, 2012, 3509–3511.
84. T.-L. Wee, L. C. Schmidt and J. C. Scaiano, *J. Phys. Chem. C* **116**, 2012, 24373–24379.
85. A. Tanaka, Y. Nishino, S. Sakaguchi, T. Yoshikawa, K. Imamura, K. Hashimoto and H. Kominami, *Chem. Commun.* **49**, 2013, 2551–2553.
86. X. B. Ke, X. G. Zhang, J. Zhao, S. Sarina, J. Barry and H. Y. Zhu, *Green Chem.* **15**, 2013, 236–244.
87. A. Pineda, L. Gomez, A. M. Balu, V. Sebastian, M. Ojeda, M. Arruebo, A. A. Romero, J. Santamaria and R. Luque, *Green Chem.* **15**, 2013, 2043–2049.
88. A. Tanaka, K. Fuku, T. Nishi, K. Hashimoto and H. Kominami, *J. Phys. Chem. C* **117**, 2013, 16983–16989.
89. A. Marimuthu, J. W. Zhang and S. Linic, *Science* **339**, 2013, 1590–1593.
90. Q. Xiao, S. Sarina, A. Bo, J. F. Jia, H. W. Liu, D. P. Arnold, Y. M. Huang, H. Wu and H. Y. Zhu, *ACS Catal.* **4**, 2014, 1725–1734.
91. Y. L. Zhang, Q. Xiao, Y. S. Bao, Y. J. Zhang, S. Bottle, S. Sarina, B. Zhaorigetu and H. Y. Zhu, *J. Phys. Chem. C* **118**, 2014, 19062–19069.
92. X. Y. Zhao, R. Long, D. Liu, B. B. Luo and Y. J. Xiong, *J. Mater. Chem. A* **3**, 2015, 9390–9394.
93. T. Mitsudome, A. Noujima, Y. Mikami, T. Mizugaki, K. Jitsukawa and K. Kaneda, *Angew. Chem. Int. Edit.* **49**, 2010, 5545–5548.
94. J. Ni, L. He, Y.-M. Liu, Y. Cao, H.-Y. He and K.-N. Fan, *Chem. Commun.* **47**, 2011, 812–814.
95. F.-Z. Su, L. He, J. Ni, Y. Cao, H.-Y. He and K.-N. Fan, *Chem. Commun.* 2008, 3531–533.
96. Y. Zhu, H. F. Qian, B. A. Drake and R. C. Jin, *Angew. Chem. Int. Edit.* **49**, 2010, 1295–1298.
97. E. Merino, *Chem. Soc. Rev.* **40**, 2011, 3835–3853.

Chapter 7

Harnessing Nature's Purple Solar Panels for Photoenergy Conversion

Elena A. Rozhkova* and Peng Wang[†]

*Center for Nanoscale Materials, Argonne National Laboratory,
9700 South Cass Ave, Argonne, IL, 60439-4855, USA
[†]State Key Laboratory of Crystal Materials, Shandong University,
Jinan, Shandong 250100, PR China

1. Introduction

Nanophotocatalysis is one of the potentially efficient ways of solar energy conversion. Translation of global power demand into novel emerging technologies could predictably lead to the cost of solar power drop below retail electricity in the next few years.[1] Being naturally free, inexhaustible and environmentally benign, solar power could become the world's dominant source of electricity by the mid-century, ahead of fossil fuels, wind, hydro and nuclear (Figure 1).[2-4]

Transformation of solar energy into heat, electricity, or chemical energy has been a motivation of many successful attempts.[5] Solar energy can be converted directly into a clean chemical fuel hydrogen H_2[6-8] via photocatalytic or photoelectrochemical (PEC) water splitting reaction.

---

*Corresponding author: rozhkova@anl.gov

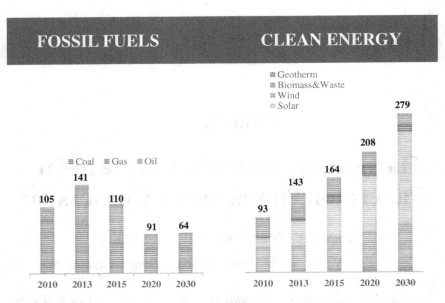

Fig. 1. Energy production forecast (in gigawatts) in new plants that burn fossil fuels (left) and utilize of renewable electricity capacity (right).[2]

When combined with oxygen in a fuel cell it produces a heat and electricity and water as a by-product:

$$2H_2 + O_2 \rightarrow 2H_2O \qquad (1)$$

Thermodynamically, H_2 combustion can provide triple the energy of gasoline or natural gas (per gram).[5]

In the natural world conversion of sunlight to chemical energy is carried by *phototrophs*, organisms that capture and utilize sunlight photons to produce energy-storage organic molecules. Biological energy conversion is accomplished via two evolutionary-independent schemes by direct translocation of protons (using rhodopsins) or electrons (using photosynthetic reaction centers) across a membrane.[9]

The overall photosynthetic reaction includes two coupled processes — oxidation of water, also known as "oxidative water splitting", which requires photons and generates electrons and protons (2) and light-independent step of CO_2 reduction to glucose (3):

$$2H_2O + h\nu \rightarrow O_2 + 4H^+ + 4e^- \qquad (2)$$

$$CO_2 + 4\,H^+ + 4e^- \rightarrow n\,(CH_2O) + H_2O \quad (3)$$

This process of solar energy conversion known as photosynthesis is one of the Earth's life fundamentals.[9] Principles of biological photoenergy conversion has been an inspiration for the development of artificial energy conversion, known as *artificial photosynthesis,* for many years.[10–12] Artificial photosynthetic schemes allow for capturing and transformation of energy of sunlight into chemical fuels via mechanisms conceptually very similar to biological conversion of light and water into carbohydrates and oxygen. In artificial bio-inspired systems reactions (2) and (3) are catalyzed by engineered photocatalysts and performed separately.

The oxidation-reduction reaction (1) includes two half-reactions

$$H_2O + 2h^+ \rightarrow 2H^+ + \tfrac{1}{2}\,O_2 \text{ (oxidation half-reaction)} \quad (4)$$

$$H^+ + e^- \rightarrow \tfrac{1}{2}\,H_2 \text{ (reduction half-reaction)} \quad (5)$$

The reaction (4) is also known as oxygen evolution reaction (OER), while reaction (5) that leads to hydrogen evolution is often called HER. Efficient, robust and low-cost catalysts of these reactions are among key components of forthcoming clean energy technologies. An ideal artificial photosynthesis catalyst is expected to drive a process that evolves both oxygen and hydrogen. However most efforts of scientific teams are still focused on studying the separate half reactions while examples of overall water splitting remain limited.[13] Sacrificial electron donors (hole scavengers) or acceptors are utilized to decouple HER and OER, respectively, in experimental settings.

Nature, besides inspiring scientists and engineers conceptually, also provides biomolecules, complexes and machineries which owing their inherent functionality can be employed as backbones for engineering of advanced bio-inspired hybrid photocatalysts.[14–18]

With the advent of innovative nanotechnology and nanomaterials remarkable leap forward in the field of the artificial photosynthesis is expected. Being intrinsically nano-sized, biologically occurring energy complexes can be fused with inexpensive and stable man-made inorganic semiconductors resulting in robust dye-sensitized photocatalytic architectures with advanced tailored properties.

Based on fundamental evolutionary-distinct mechanisms of solar energy transformation in nature, biological molecules and complexes used as backbones in bio-inorganic hybrids can be broadly combined in two main groups: the first one utilizes chlorophyll-related antenna "super"-complexes (including either photosystem PS or related tetrapyrrole structures) and the second group is based on simple retinal-containing transmembrane proton pump bacteriorhodopsin (bR). In this article we initially discuss natural mechanisms of sunlight energy transformation with emphasis on rhodopsin machinery and then focus on application of bR proton pump and its membrane complex known as purple membranes in engineered artificial photosynthesis systems for energy conversion.

2. Nature's Purple Solar Panels

Natural phototrophic systems acquire sunlight energy to convert it to chemical compounds through *chemiosmosis*, the movement of ions across a membrane. The resulting electrochemical gradient fuels ATP-synthases to produce ATP, a key biological energy carrier. This universal process is accomplished via two evolutionary distinct mechanisms.[9] The first one, a multistep electron-shuttling mechanism, photosynthesis that utilizes green pigment chlorophyll has evolved at the earliest period of earth history, approximately 2.4 billion years ago. Chlorophylls have been identified in complex of proteins, pigments and cofactors of the reaction centers (RC) in plants, algae and cyanobacteria. Structurally chlorophyll pigments represent magnesium ion-centered circulated tetrapyrrole complexes, see Figure 2, which provide RCs charge separation capability and, in addition, serve as principal antennae for light capturing.[9] Some cyanobacteria also utilize bilins, open chain tetrapyrrole pigments, Figure 2, as accessory antennae which assist chlorophylls in capturing more photons and funneling light energy to the RC. Both cycled[18] and linear[15,16,19] natural tetrapyrrole complexes have been successfully utilized in artificial photosynthesys schemes.

Compared to chlorophyll-based multicomponent RCs, rhodopsins utilize solar energy via alternative amazingly simple mode based on sunlight-driven proton transfer across the membrane by a proton pump bacteriorhodopsin.[20] The latter mechanism and the archetypal rhodopsin were discovered by D. Oesterhelt and W. Stoeckenius in Archaea, *Halobacterium halobium*.[21,22] The light-sensitivity of rhodopsins and their color result

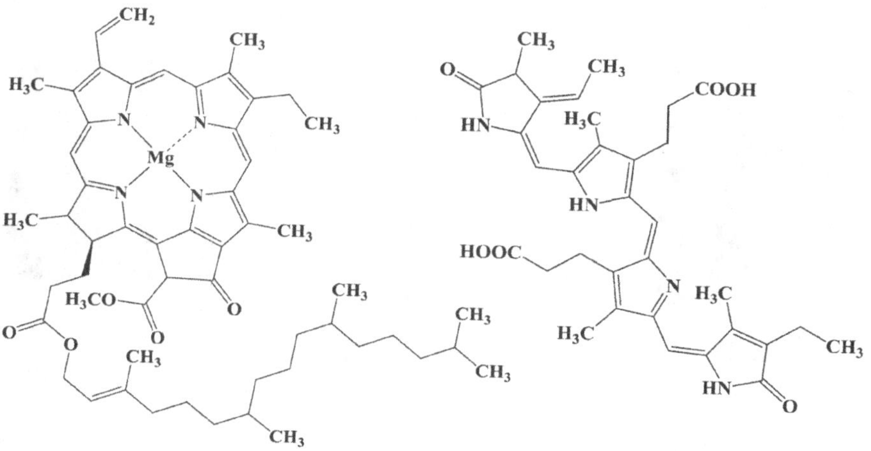

Fig. 2. Chemical structure of tetrapyrroles chlorophyll (left) and phycobilin (right).

from the presence of the carotenoid retinal, a chromophore group covalently linked to opsin apoprotein, Figure 3(b). Besides photoenergy conversion in lower organisms rhodopsin photoreceptors (related to bR) also serve as visual pigments involved into the intra- or intercellular signaling in higher animal vision and the circadian clock mechanisms.[20] Despite of differences in sequences, microbial and animal rhodopsin share a common tertiary architecture with the N- and C-terminus facing out- and inside of a cell membrane.[20]

bR proton pumps are relatively small 26 kDa protein architectures which can be considered as prototype membrane transporting devices capable of carrying ions against an electrochemical potential — up to 250 millivolts, which translates into a 10,000-fold difference in proton concentration on either side of the membrane.[23] The resultant electrochemical gradient is further utilized in the ATP synthase-driven phosphorylation that produces adenosine triphosphate (ATP), the main fuel necessary to sustain cell's life, Figure 3(d).

The bR pumps are neatly arranged as 2-D nanocrystal lattice integrated into the bacterial cell membrane with a uniform orientation which is known as the purple membrane (PM). Being expressed by an extremophile organisms Archaea that thrives in harsh "sterile" environments such

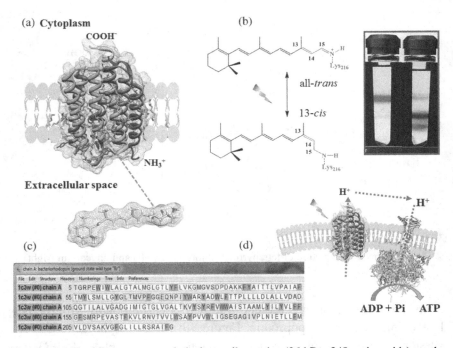

Fig. 3. (a) The bR pumps are relatively small proteins (26 kDa, 248 aminoacids) neatly arranged as a two-dimensional (2D) nanocrystal lattice integrated into the bacterial cell membrane with a uniform orientation that is known as the purple membrane (PM). (b) The light-sensitive retinal cofactor, the aldehyde of vitamin A, is covalently linked to Lys 216 (K 216) forming *protonated* Schiff base (–HC=NH$^+$–). Inset: Isolation of PMs (lower purple bands) from Halobacteria via sucrose gradient fractioning. (c) The protein sequence includes multiple aromatic amino acids: 8 tryptophan (W), 11 tyrosine (Y) and 13 phenylalanine (F) residues out of 226 amino acid residues, highlighted in dark green. Y83, W86 and W182 are strongly conserved through microbial rhodopsins sequences. (d) The transmembrane proton gradient created by bacteriorhodopsin photoconverter is then used in ATP synthase-driven phosphorylation.

as salt lakes and hot springs where no other microorganism can survive, PMs are naturally evolved to tolerate and maintain their photoreactivity under demanding conditions, including exposure to light and oxygen, high ionic strength, temperatures and broad range of pH values, as summarized in the Table 1. Stability of the PMs arises from specific features of an archaeal membrane structure. Thus, chemically transmembrane lipids of

Table 1: PMs as a robust natural material with outstanding properties (adapted from Refs. 26–28)

Physicochemical properties of PMs
Composition: lipids and bR protein (26 kDa) at a molar ratio of about 3:1 arranged in a 2D hexagonal crystalline lattice
Chromophore: all-trans retinal, vitamin A aldehyde, bound stoichiometrically bound via a Schiff base linkage to the amino group of a lysine side chain
Refractive index: n_{PM} = 1.45–1.55
Characteristic absorbance at λ_{max} 570 nm, ε = 63,000 M^{-1} cm^{-1}
Thickness: 5 nm
Buoyant density: 1.18 g/cm^3
Stable: under sunlight exposure in the presence of oxygen for years — at temperature over 80°C (in water) and up to 140°C (dry) — at pH values from 0 to 12 — high ionic strength (3 M NaCl) — preserve color and photoactivity when dried — stable in non-polar solvents (e.g. hexanes)
Can be digested by most proteases. Sensitive to polar solvents

the PM represent L-glycerol with inverted stereochemistry (L- instead of D-glycerol in bacteria and eukaryotes) linked to branched isoprene chains (in contrast to easier oxidized unbranched fatty acids in other organisms) via more stable to hydrolysis ether bond.

Halophilic (salt-loving) Archaea microorganisms employ PM as a "purple solar panels" to capture sunlight energy and to convert it into chemical energy of ATP and power the cell. The transmembrane bR protein consists of seven α-helices A-G (Figure 4) burying a retinal chromophore covalently linked to Lys 216 in a helix G via protonated Schiff base linkage, Figures 3(a) and (b). The retinal group is located approximately in the center of the PM, at a distance of nearly 2.5 nm from both PM surfaces.[24]

The bR photocycle is initiated upon photon absorption by the chromophore and results in vectorial translocation of proton across a membrane from the cytoplasm to the extracellular side. Retinal undergoes subpicosecond photoisomerization from all-trans (the ground state) to 13-cis in response to visible light, λ max ~570 nm, as shown in Figures 3(b) and 4.

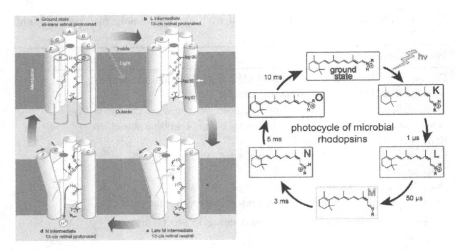

Fig. 4. Molecular mechanisms of proton (H⁺) pumping in bacteriorhodopsin includes ground state and intermediates which exhibit specific spectral characteristics. Light-induced isomerization of the protonated retinal from all-*trans* (purple) to 13-*cis* (pink) triggers the transfer of the proton through a network of hydrogen bonds of the transmembrane protein toward the extracellular space. [With permission from Nature 2000[23] and from ACS[20].]

Owing to well-defined chromophore and aminoacids framework layout the protonated Schiff base (–HC=NH+) then release the proton (possibly, to aspartic acid residue Asp85) followed by alterations in protonatable groups within the apo-protein molecule to the extracellular side. The photocycle undergoes through spectrally distinguishable intermediates named as K, L, M, N and O, where K and O are the red-shifted intermediates, while L, M, and N are all blue-shifted intermediates, Figure 4. The Schiff base deprotonates upon intermediate M formation and then re-protonate upon M decay. Millisecond re-isomerization to all-trans conformation occurs upon O intermediate formation which then slowly relaxes thermally to the ground state. In overall, per one photon of light absorbed two protons H⁺ are translocated through a membrane. The quantum efficiency (QE) of bR photo conversion was determined to be 64%.[25] It is important to re-emphasize in here that comparing to chlorophyll-based PS there is no charge separation and no electron transfer involved into bR-mediated sunlight energy transformation. While electron transfer through bR protein was successfully demonstrated experimentally in model lipid bilayers arranged on electrically conducting supports,[26] it is still intriguingly unclear why this function is not realized in nature.

PMs can be routinely isolated from halophilic bacteria via simple cell lysis by exposing cells to low salt or distilled water followed by sucrose gradient centrifuge fractioning, Figure 3, inset.[29] Owing to inherent structural features and exceptional properties of the nature's "purple solar panels" along with established methods of their isolation and purification from a low maintenance microorganism PMs represent an excellent biomaterial for fabrication of green energy hybrid devices.

3. Recent Developments in Engineered bR and PM-based Hybrid Systems for Photoenergy Application

Thermodynamically the water splitting is an "uphill" endothermic reaction with a positive change in Gibbs free energy of $\Delta G^0 = 237$ kJ mol^{-1}. The photon energy is used to overcome the large positive change in Gibbs energy and shift the reaction equilibrium toward water splitting. The first successful example of "artificial" splitting of water (photolysis) to oxygen and hydrogen using a semiconductor TiO_2 photoanode and platinum counter electrode with an external bias under UV light irradiation was reported by A. Fujishima and K. Honda in 1972.[30] In their photoelectrochemical cell (PEC) photo-excitation of TiO_2 photoanode, an n-type semiconductor, by light with energy greater than the band gap of TiO_2, results in generation of separated electrons and holes in the conduction band and valence band, correspondingly. The water molecule is then oxidized by the positive photogenerated hole producing oxygen, while the proton (hydrogen ion) is reduced by the photogenerated electrons on platinum electrode evolving hydrogen, as presented in equations (2) and (4) above.

Since this breakthrough discovery there has been a great continuous interest of scientists and engineers in developing of visible-light-responsive photocatalysts because UV light accounts for only ≤4% of solar photons. Various methods of extending the visible light reactivity of semiconductor include doping[31,32] and sensitization by organic and metalorganic dyes[33] or, more recently, quantum dots.[34] On the other hand, biological materials, including photosynthetic components,[15,18] enzymes[17] and biomass derivatives[35] were also proposed as building blocks in engineered environmentally-friendly visible light catalytic systems. While chlorophyll-related natural complexes, such as PSs, are sensitive to environment when isolated from their natural source and require multiple mediators and cofactors, purple membranes containing light-driven proton pump bacteriorhodopsin are very attractive materials for engineering artificial hybrid systems for solar energy transformation owing to their structural simplicity and elegance, excellent photochromic properties and robustness. Photovoltaic characteristics of bacteriorhodopsin were reported by Li and coauthors.[36] The energy levels of the lowest unoccupied molecular orbital (LUMO) and highest occupied molecular orbital (HOMO) for bR were identified as −3.8 eV and −5.4 eV, respectively, consequently resulting in HOMO-LUMO gap of 1.6 eV. When bR chromophore molecule is integrated with TiO_2 which conduction energy is −4.2 eV, the photoexcited electron from bR can be injected into a lower energy conduction band of the semiconductor, Figure 5.

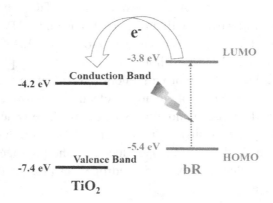

Fig. 5. Energy level diagram and proposed photoinduced charge carrier injection within bR-TiO_2 hybrid. While LUMO/HOMO of bR were determined to be −3.8 eV and −5.4 eV, respectively, the CB energy of TiO_2 located at −4.2 eV. This difference makes electron injection from the bR into semiconductor energetically favorable.

Photocatalysts for artificial photosynthesis can be utilized through two principal schemes. First, photocatalyst is a key element in PEC designed for producing electrical energy or hydrogen via Honda-Fujishima's[30] water electrolysis. In the second simpler, cost efficient and scalable photocatalysis scheme anode and cathode materials are integrated without electric circuit as a unified hybrid photocatalyst which is suspended in aqueous solution.[37,38] Advancing both PEC and photocatalysis schemes to visible light-driven water splitting can be implemented via Gratzel's dye-sensitizing of a wide-gap semiconductor[39–41] by organic molecules or by inorganic narrow band gap semiconductor nanocrystals (e.g. QDs). Synthesizers are capable of harvesting visible light and injecting photoexcited electrons into the conduction band of a semiconductor. In the following section, we first overview successful reports on hybrid PEC and then we focus on engineered particulate photocatalytic systems based on bacteriorhodopsin and Purple Membranes.

3.1 PM-based photoelectrodes

bR and PMs have been integrated with semiconductor photoelectrodes to facilitate absorption of photon energy of the visible light in dye-sensitized solar cells by few research groups.[42–46] Thus, Allam and co-authors reported on fabrication of photoanode material through assembly of bR on ~7 µmTiO$_2$ nanotubes array, Figure 6.[42] Under AM 1.5 illumination (100 mW/cm^2) the hybrid electrodes achieved a photocurrent density of 0.65 mA/cm^2 which is a ~50% increase over that measured for pure TiO$_2$ nanotubes (0.43 mA/cm^2) under the same conditions, while in the presence of I$^-$/I$_3^-$ as a redox electrolyte the photocurrent increased to 0.87 mA/cm^2. The process of PEC water-splitting reaction using TiO$_2$-bR photoanode can be described by equations below:

TiO$_2$ + hv → e$^-$ + h$^+$ (6)

Anode: H$_2$O + 2h$^+$ → 2H$^+$ + ½ O$_2$ (4)

Cathode: 2H$^+$ + 2 e$^-$ → H$_2$ (5)

Overall: 2 hv + H$_2$O → ½ O$_2$ + H$_2$ (2)

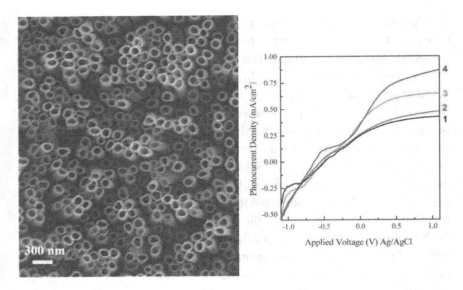

Fig. 6. bR/TiO$_2$ nanotube arrays hybrid for PEC water splitting. Left. Field emission scanning electron micrograph of the fabricated TiO$_2$ nanotube arrays (top view). Right. Photocurrent density versus potential under AM 1.5 G illumination (100 mW/cm^2) of pure TiO$_2$ nanotubes (1), bR/TiO$_2$ with 500–600 nm light cut off (2), bR/TiO$_2$ (3), and bR/TiO$_2$ with redox molecules in the electrolyte (4). [Reproduced with permission from the Energy & Environmental Science, RCS[42].]

Absorption of photons by the semiconductor results in the charge separation and the formation of electron-hole pair (e$^-$/h$^+$). The water is then oxidized by the photogenerated holes, while the protons are reduced by the electrons at the cathode. Similarly to other bio-synthesized solar cells, for example earlier reported chlorophyll-based systems developed by Gratzel,[10,33,39–41,47] photoexcited bR functions in this system as a dye that injects electrons into the conduction band of TiO$_2$ array electrode thus extending its visible light absorption capability. Moreover, in the absence of light that coincides with bR absorption band (λ incident 500–600 nm) the performance of the bR/TiO$_2$ photoanode was the same as that of pure TiO$_2$ indicating that proton pumping provides "an additional source" of photocurrent enhancement, Figure 6.[42] Finally, the authors compared two methods of assembly bR molecules on the electrode — random self-assembly and covalent anchoring using 3-Mercaptopropionate linker. The latter method resulted in higher stability of the electrode-biomolecule assembly. Thus, after several electrochemical runs remarkable amounts of bR self-assembled on the electrode with no linker were dissociated into the electrolyte.[42]

Later Naseri and others likewise reported on bR-nanotube TiO_2 photoanodes array[43] proposing original non-covalent grafting approach through immersion of just-annealed hot semiconductor films into a bR solution. Such chemical linker-free approach could be attained owing to excellent thermal stability of the PMs, but most possibly, it cannot be applied for other common organic dyes and biomolecules which are much more sensitive to high temperatures. Besides alternative protein engineering tactics have been utilized to enhance bR-electrode interface. For example, Thavasi and coauthors altered three glutamic acid residues located in the extracellular region of bR protein to glutamine residues (E9Q/E194Q/E204Q) and demonstrated theoretically and experimentally that these alterations can facilitate more favorable interfaces between bR and oxygen atoms of TiO_2 anatase via electrostatic interactions.[44] A solar cell composed by the triple mutant demonstrated higher photoelectric response than that included the wild type bR.[44] Another Cys 247 bR mutant engineered by Renugopalakrishnan and colleagues allowed for controllable well-ordered covalent attachment of the bR to gold shell composite electrode.[45]

3.2 Nanomaterials-enabled bR's photocycle enhancement

bR absorbs green light (λ_{max} at 560 nm) due to presence of the retinol cofactor that forms a protonated Schiff base with the Lys 216 (K 216). Reversible protonation of the Schiff base is the key of the photoreception mechanism of bR. Apparently, extending light absorption properties of a biological structures is possible using powerful arsenal of photonic NPs. The intrinsic photophysical properties of bR and PMs were expanded using various photonic NPs towards energy conversion applications. For example, it was demonstrated that Au, Ag and Au-Ag alloy NPs notably enhanced the proton pumping capability of the bR under the NPs plasmonic field via altering kinetics of the protein's natural photocycle.[48–52] Recently, Zhao et al. developed a new bio-hybrid cathode for electrocatalytic hydrogen production via water splitting.[53] The cathode composed of silver NPs and bR supported on solid carbon cloth (CP) showed a low onset overpotential of 63 mV and good stability (up to 1000 cycles) in alkaline media.[53]

From the other hand, natural light harvesting function of bR machinery can be extended via tagging of the PMs with CdTe or CdSe/ZnS quantum dots (QDs) that allows for harvesting light from deep-UV to blue region through Forster resonance energy transfer (FRET) from QDs to the biomolecules.[45,54] The coupling of QDs (donor) and bR retinal (acceptor) allowed for achieving FRET with nearly 100% efficiency. Moreover, it was demonstrated that the bR being integrated into the engineered QD-PM hybrid material demonstrated remarkable (~25%) increase of the photoresponse for the proton pumping.[54]

Whereas the solar spectrum includes only 4% of UV, 44% visible and over 50% of infrared (IR) light, opportunities to tune intrinsic photophysical properties of "host" materials toward IR wavelength is very attractive. Lu and co-authors[55] reported on integration of PMs with upconversion polyethylenimine (PEI)-modified NaYF4:Yb (20%), Er (2%) nanocrystals that allowed for producing of the NIR-triggered photoelectrochemical responses. Excitation of these upconversion nanoparticles by NIR light (980 nm, 2000 mW cm^{-2}) results in three emission peaks at 524, 543 and 658 nm. The strongest emission of the upconversion NPs is in the wavelength range of 520 to 580 nm that overlaps with the maximum absorption of the bR. Therefore, the strong photoluminescence of the particles provides a source of trigger of the bR photocycle. This work demonstrated further potential of using bR in the NIR wavelength range.

3.3 bR-based nanodevices for sunlight energy conversion

Natural proton pumps can also be employed as functional elements in multicomponent sophisticated bio-inspired energy nanoconstructs. For example, bR and related retinylidene proton pump with conserved key structural and mechanical features proteorhodopsin (pR), were utilized by

Xiang's group to design various light-driven energy bio-devices. Recently the group constructed a multilayer heterogeneous photovoltaic stack device where solid layers of bR served as photon acceptors while Au NPs enhanced the photocurrent owing to the surface plasmonic effect and shortening of the bR natural photocycle. Such stack structure is inspired by the naturally-occurring 3-dimentional (3-D) granum network in chloroplast. The photocurrent recorded in this integrated stack system reached to 350 nA cm^{-2} under optimized conditions.[56] Further, inspired by fundamental capability of plasma membrane of charging and discharging the group constructed a light-powered bio-capacitor. While the biological prototype capacitor-like behavior is enabled by cooperation between membrane proton pump and channel architectures, Xiang's device includes natural light-inducible pump pR and a fabricated "surrogate" alumina nanochannel with tunable diameter.[57] In this device the proton pump serves for photocurrent generation whereas the manmade nanochannel acts as an adjustable resistance. Capacitor-like behavior of the biomimicking hybrid system was demonstrated through observation of constant-ratio relationship between the photocurrent duration time and the nanochannel resistance.[57]

3.4 Photocatalytic systems for photon-to-hydrogen conversion based on TiO$_2$ and PM

Efforts of our group have been focused on employing the light-harvesting proton pump bR as a building block to construct environmentally benign, robust, cost efficient and scalable *"wireless"* photocatalytic particulate (slurry) systems for visible light-driven hydrogen production.[58,59] Principles of the proposed solar-driven nano-bio assembly are summarized in the Figure 7.[58] In this scheme the light-harvesting proton pump bR self-assembled on the surface on TiO$_2$ is utilized as a backbone that similarly to traditional organic dye photosynthesizers assists for harnessing the visible light reactivity of TiO$_2$ photocatalyst.

Besides, bR also retains its biological function of "harvesting" and driving protons toward platinum NPs co-catalyst for consequent reduction to hydrogen gas. TiO$_2$ NPs provide charge separation functionality. Absorption of photons by the semiconductor NPs leads to the charge

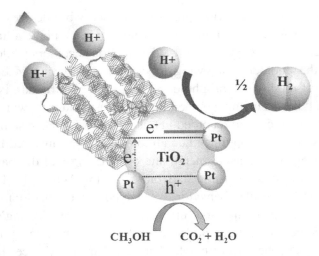

Fig. 7. Principle scheme of solar H_2 generation catalyzed by nano-bio hybrid assembly of bacteriorhodopsin (bR) and Pt NPs co-catalyst on TiO_2 under visible light. [Reproduced with permission from ACS[58].]

separation and the formation of electron-hole pair through the equation (6). The photoinduced electrons are then consumed for the protons reduction over co-catalyst Pt NPs, while excess of holes is scavenged by a sacrificial electron donor methanol.

Pt co-catalyst nanoparticles with narrow size distribution (~4 nm) were deposited on the surface of TiO_2 P25 clusters via photoreduction of a hexachloroplatinate precursor in the presence of a sacrificial electron donor. Next, PMs self-assembled on the TiO_2 nanoparticle owing to the presence of several charged residues on the cytoplasmic side as well as on the extracellular side, Figure 3. Specifically, the carboxylic groups presented on the cytoplasmic side can allow for predominant self-assembling of bR molecules on the exposed surface oxygen atoms of anatase NPs. The formation of stable conjugate of PMs with TiO_2 was confirmed by Raman microscopy. Raman spectra of the nano-bio assembly were dominated by a broad ethylene-stretching mode of the retinal chromophore at 1530 cm^{-1} and fingerprint modes at 1171 and 1200 cm^{-1} for the bR molecule as well as with the fingerprint modes of the anatase TiO_2.

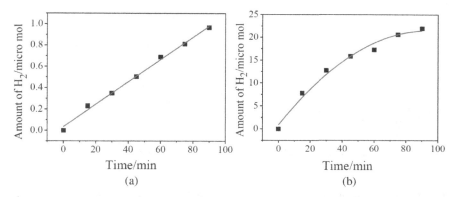

Fig. 8. Photocatalytic H_2 production in the presence of 25 mM methanol as electron donor at pH7 (a) under monochromatic green light, 560 ± 10 nm (13 mW/cm^2) and (b) under white light illumination (350 nm ≤ λ ≤ 800 nm, 120 mW/cm^2). [Reproduced with permission from ACS[58].]

When aqueous slurry of the nano-bio photocatalyst was exposed to green light (560 ± 10 nm), continuous H_2 generation was observed with a turnover rate of 207 μmole of H_2 (μmole protein)$^{-1}$ h^{-1}, Figure 8a. On average, under light illumination nearly constant H_2 evolution by the nano-bio catalyst has been observed at least for 2.5 h. Control experiments with Pt/TiO$_2$ without bR yielded in minute amount of H_2 evolved for 1 h of illumination with green light. When monochromatic light was replaced with white light illumination with λ = 350–800 nm, H_2 turnover rate was increased by 25 times to 5275 μmole of H_2 (μmole protein)$^{-1}$ h^{-1}, Figure 8b, most possibly due to additional electrons originated from excitation of TiO$_2$ nanoparticles. Thus, the Pt/TiO$_2$/bR hybrid photocatalyst outperforms many other reported nano-bio assemblies in photocatalytic hydrogen generation.

In order to establish the role of bacteriorhodopsin in the above hybrid system, photoelectrochemical measurements were carried out. bR-modified TiO$_2$/FTO electrodes were tested for the photocurrent generation with white light as well as under monochromatic green light illumination (560 nm). Figure 9 shows the linear sweep voltammograms measured on bR-modified TiO$_2$/FTO electrodes under dark and under white light

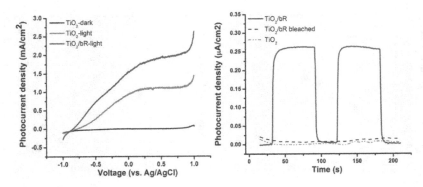

Fig. 9. I–V characteristics of TiO2 and TiO2/bR photoelectrode in dark and light illumination (100 mW/cm²), Left, and short-circuit photocurrent response of TiO_2, TiO_2/bR, and TiO_2/bR hydroxylamine bleached electrode in aqueous electrolyte containing 5 mM hydroquinone (HQ) in 10 mM MES, pH 6.2 buffer. [Reproduced with permission from ACS[58]]

illumination. In dark, both the electrodes show negligible current response as the potential was scanned from −1.0 to 0.9 V (vs Ag/AgCl). Upon illumination, the bare TiO_2 electrode has an onset photocurrent at −0.75 V and the current continues to increase before reaching a steady state value of 1.0 mA/cm². When TiO_2 photoelectrode was modified with bR, the steady-state photocurrent density value increased to 2.0 mA/cm² that was ~50% over pure TiO_2 electrode.

Further, under the green light illumination (λ = 560 nm) the photocurrent increased rapidly and reached a steady-state current density of 0.27 µA/cm². The photocurrent returned to the background level after illumination was turned off and the process could be repeated several times. Conversely, unmodified TiO_2 electrode and TiO_2 electrode modified with bleached bR chromophore showed no current response when illuminated with green light demonstrating that the origin of the photocurrent could be attributed to the excitation of bR retinal chromophore and associated charge transfer to the TiO_2 nanoparticles. Transient absorption (TA) measurements further verified charge transfer from bR molecule to TiO_2 nanoparticle. TA measurements were carried out using a pump wavelength of 560 nm such that it excited only the bR molecule leading to a charge injection to TiO_2. The bleach of the excited bR molecule, monitored at

625 nm, was examined in the presence of different concentrations of TiO_2 nanoparticles (0, 20, 100 μg ml^{-1}). Figure 10 shows the lifetime decay trend for the bR molecule with and without TiO_2 nanoparticles. For bR molecule, the excited state decays with a time constant of 1.03 ± 0.11 ps whereas for bR/TiO_2 the decay lifetime was 0.42 ± 0.11 and 0.16 ± 0.02 ps for 20 and 100 μg ml^{-1} TiO_2 nanoparticle concentration, correspondingly. These sub-picosecond bleach recovery dynamics indicate a charge separation efficiency from excited bR to TiO_2 of >85%.

In is noteworthy that in comparison to other reports on utilizing bR in PECs for our photocatalytic system (particles slurry) we did not observe considerable improvement of a photocatalyst performance and stability by means of covalent attachment of bR to TiO_2 nanoclusters.[58] Utilizing biologically-inspired self-assembly approaches for engineering nano-bio hybrids with using no coupling chemicals is attractive due to greener environmentally friendly character and lower cost. Further we proposed to introduce an additional component — reduced graphene oxide (rGO) — to the bR-TiO_2-Pt assembly that allowed for seamless interfacing and

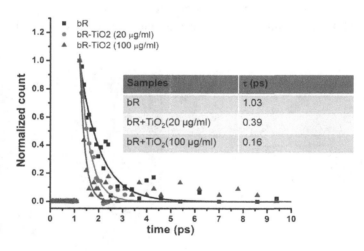

Fig. 10. Ultrafast transient absorption measurements showing the lifetime decay of the excited bR species reveal charge transfer from bR molecules to TiO_2 particles. The samples were pumped using 560 nm and probed at 625 nm laser pulses. Kinetic traces at 625 nm of bR and bR-TiO_2 samples along with exponential fit. [Reproduced with permission from ACS[58].]

promoting interactions between the biological molecules and nanoparticles as well as between distinct TiO_2 nanoclusters.[59]

Further performance boosting of bR-based hybrid photocatalyst was realized via incorporation of an additional material — atomically thin carbon film graphene.[59] Graphene, a single layer of sp^2-bonded carbon atoms arranged in a two-dimensional honeycomb lattice, possesses excellent mechanical, thermal, and optical characteristics, high conductivity, and electron mobility properties and provides large specific surface area.[60] Hydrophilic graphene oxide (GO) and reduced GO (rGO) represent graphene functionalized with oxygen-centered groups for instance carboxylic, hydroxyl, or epoxide. Presence of these functional groups allows for interaction with positively charged biomolecules, including bR, and can serve as nucleation centers for nanocrystals growing. As sketched in Figure 11, the designed a nano-bio photocatalyst that employs both reduced graphene oxide and membrane proton pump (bR) lattices as building blocks for sensitizing the TiO_2 photocatalyst to visible light, while platinum nanoparticle co-catalyst utilizes photoinduced electrons to reduce protons to hydrogen. Photoinduced holes are scavenged by a sacrificial substrate.

The Pt/TiO_2-rGO system was fabricated *via* simultaneous photoreduction of GO and sodium hexachloroplatinate precursors on TiO_2 in the

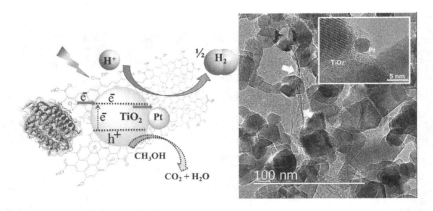

Fig. 11. Depiction of Pt/TiO_2-rGO photocatalyst hydrogen evolution under visible light (left) and TEM images of the Pt/TiO_2-rGO (right). The white arrows point to the atomically thin rGO sheets assembled on the surface of the TiO_2. Inset: HRTEM image of Pt co-catalyst nanoparticles photodeposited on TiO_2. [Reproduced with permission from ACS[59].]

presence of ethanol. The total amount of graphene material used to assemble with TiO_2 was 0.5 wt %. As it is evident from the TEM, Fig. 11, rGO flakes were uniformly dispersed through TiO_2 nanoclustres. Pt nanoparticles with an average diameter of 3 nm grown predominantely on the semiconductor surface can be observed in the HRTEM micrograph, as shown in the inset, Figure 11. Restored sp^2 networks of rGO and multiple available oxygen-centered surface groups allow for the unified connection between the carbon material and exposed oxygen atoms at the surface of TiO_2 nanoparticles. Subsequent self-assembly of membrane–bR complexes on the surface of the Pt/TiO_2-rGO resulted in the Pt/TiO_2-rGO-bR visible-light-reactive nano-bio photocatalyst. Over again, we employed the natural structural features of the bR that allows for utilizing self-assembly, a minimalistic bioinspired time- and cost-efficient approach. Owing to availability of multiple charged groups on the cytoplasmic and extracellular sides of the PM, the biological membrane complex readily interacts with oxygen-centered groups on the rGO. Furthermore, multiple aromatic amino acid residues (Thr, Phe, Tyr) exposed on extracellular surface (Figure 3) further promote well-ordered assembly of the flat 2D membrane protein on 2D rGO lattices *via* π–π stacking interactions. In addition, some of the PMs can assemble on the exposed oxygen atoms of the anatase surface of the TiO_2 nanoparticles. Hence, in aqueous solution, bR self-assembles on the rGO-modified TiO_2 to form a stable conjugate. The EPR data that provide further insights into the hybrid catalyst structure and interactions between biological and nanoparticle components will be discussed below.

Photocatalytic H_2 evolution over the Pt/TiO_2-rGO-bR nano-bio photocatalyst was evaluated under green and white light irradiation using methanol as a sacrificial electron donor. The H_2 production activities of the Pt/TiO_2-bR were also measured for reference. The amount of Pt in Pt/TiO_2-bR and Pt/TiO_2-rGO-bR were optimized experimentally. The highest photocatalytic activities for hydrogen evolution were obtained when Pt content was 0.75 wt % in Pt/TiO_2-bR and 0.5 wt % in the carbon-material-boosted counterpart Pt/TiO_2-rGO-bR. As shown in Figure 12a, under continuous green light (560 ± 10 nm) illumination, H_2 generation was observed with a turnover rate of approximately 298 μmol of H_2 (μmol protein)$^{-1}$ h^{-1}. The turnover rate of Pt/TiO_2-rGO-bR is higher

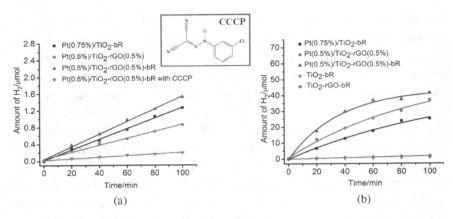

Fig. 12. Photocatalytic H_2 evolution in the presence of methanol as electron donor at pH 7.0 (a) under monochromatic green light, 560 ± 10 nm (13 mW/cm²), and (b) under white light illumination (350 nm ≤ λ ≤ 800 nm, 120 mW/cm²). Inset: structure CCCP, a protonophore and uncoupler that causes an uncoupling of proton gradient. In the case of the Pt-TiO_2-rGO-bR assembly CCCP cancels the pumping activity of the biological proton pump and also can possibly interfere with overall catalytic proton formation and reduction. [Modified with permission from ACS[59].]

than that of the rGO-free Pt/TiO_2-bR bio-nano catalyst (229 μmol of H_2 (μmol protein)$^{-1}$ h^{-1} Noteworthy, control experiments with Pt/TiO_2 and Pt/TiO_2-rGO without bR molecules yielded lower amounts of H_2 evolved under green light illumination condition. This demonstrates again contribution of PMs as important component of the photocatalytic system. When monochromatic green light was replaced with white light illumination (λ = 350–800 nm), H_2 turnover rate of Pt/TiO_2-rGO-bR was increased by 37-fold, reaching approximately 11.24 mmol of H_2 (μmol protein)$^{-1}$ h^{-1}, which was 2 times higher than that of the Pt/TiO_2-bR nano-bio catalyst alone without rGO (5.38 mmol of H_2 (μmol protein)$^{-1}$ h^{-1})m as presented in Figure 12b. This result clearly demonstrates that introduction of a small amount of rGO (0.5 wt %) leads in remarkable improvement of photocatalytic performance of the nano-bio photocatalyst in HER and also in 25% reduction of platinum content. We proposed that under white light rGO acts similarly to an organic photosynthesizer and provides additional photoexcited electrons to the conduction band of the TiO_2 particles, thus extending visible light reactivity of the semiconductor and boosting H_2

evolution over the Pt co-catalyst. Control experiments using TiO_2-bR and TiO_2-rGO-bR without Pt nanoparticles produced only minute amounts of H_2 under white light irradiation; therefore, rGO *could not* replace the Pt cocatalyst in the H_2 evolution process. Measurements of the photocatalytic activity of the Pt/TiO_2-rGO without bR were also carried out under white light illumination and resulted in lower HER rate, as shown in Figure 12b, indicating that bR can enhance the H_2 evolution process. Hence, the highest performance was exhibited by the ***ternary*** bio-hybrid photocatalyst assembly.

In order to confirm the collective impact of rGO and bR on the hybrid catalyst performance, photoelectrochemical measurements were carried out using the typical three-electrode system in a 0.1 M Na_2SO_4 electrolyte (pH 6.5). TiO_2 and TiO_2-rGO were also tested as references. All obtained electrodes were tested for the photocurrent generation under white light and green light illumination (560 ± 10 nm), as shown in Figure 13. The current–potential (I–V) curves of TiO_2, TiO_2-rGO, and TiO_2-rGO-bR electrodes under white light illumination are presented in Figure 14a. The photocurrent of the bare TiO_2 electrode continuously increases before reaching a steady state value of 8.4 $\mu A/cm^2$, while the current of TiO_2-rGO continues to increase before reaching a steady state value of 55 $\mu A/cm^2$ at 1.0 V *vs* Ag/AgCl. In the case of the TiO_2-rGO-bR photoelectrode, the photocurrent continues to increase before reaching a steady state value of 78 $\mu A/cm^2$, which is about 9-fold higher than that of the bare TiO_2 electrode. This result indicates that the photocurrent is predominantly coming from the photoexcited rGO which transfers the photogenerated electrons to TiO_2 nanoparticles. Comparing the current–potential curves of TiO_2-rGO and TiO2-rGO-bR presented in Figure 13a, one can observe that the so-called flat-band potential of TiO_2-rGO-bR (−0.306 V *vs* Ag/AgCl) is more anodical than that of TiO_2-rGO (−0.319 V *vs* Ag/AgCl). This anodical shift can be attributed to the protons pumped from the excited bR. TiO_2-rGO works as an n-type semiconductor electrode, and when its surface contacts an aqueous electrolyte, proton dissociation equilibrium on the surface is established for the surface hydroxide, as described below:

TiO_2-rGO-OH $\leftrightarrow$ TiO_2-rGO$^-$ + H$^+$

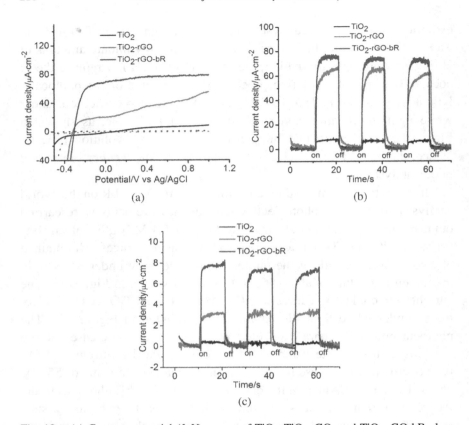

Fig. 13. (a) Current–potential (*I–V*) curves of TiO_2, TiO_2-rGO, and TiO_2-rGO-bR electrodes under white light illumination. Dotted curves correspond to TiO_2, TiO_2-rGO, and TiO_2-rGO-bR in dark conditions. (b) Photocurrent–transient responses under white light irradiation. (c) Photocurrent–transient responses under green light irradiation. TiO_2 nanoparticles and TiO_2-rGO were electrophoretically deposited on ITO electrodes to form a uniform film. The obtained TiO_2-rGO electrode was immersed in the bR solution overnight at room temperature, allowing the biomolecules to self-assemble on the TiO_2-rGO-modified ITO electrodes. The light density of white light is 120 mW/cm²; green light density is 13 mW/cm²; the electrolyte is 0.1 M aqueous Na_2SO_4, pH 6.5. [Reproduced with permission from ACS[59].]

Protons released from excited bR shifts the equilibrium to the left-hand side, which leads to the anodic shift of the flat-band potential, which is in agreement with work by Saga and co-authors.[61] Besides, HER under green light can be significantly suppressed by a proton uncoupler carbonylcyanide *m*-chlorophenylhydrazone (CCCP), a well-known inhibitor of rhodopsins proton pumping activities.[62,63] Therefore, in our bio-hybrid assembly bR functions as a traditional dye and, at the same time, retains its biological proton pumping function thus further promoting hydrogen evolution.

Figures 13(b),(c) show the photocurrent–transient responses under white light and green light illumination, respectively. In the dark, the photoelectrodes show negligible current response, while upon illumination, the photocurrent increases rapidly and reaches steady state values. Once the illumination is turned off, the photocurrent returns to the background level, and this process can be repeated many times. It follows from comparing the data presented in Figure 13(b),(c) that the origin of the photocurrent can be mostly credited to the excitation of rGO under white polychromatic light and the excitation of bR under green light and the associated energy transfer to the TiO_2.

In order to further explore electron transfer pathways within this nano-bio hybrid system electron in details electron paramagnetic resonance EPR measurements (Figure 14) were carried out on the individual and combined components. The Pt/TiO_2-rGO nanocomposites exhibit a strong sharp EPR signal in the dark with a line width of 1.45 G, close to that of delocalized electrons over a whole plane of the rGO, with the addition of a weak signal that corresponds to electrons delocalized at the edges/defect sites (Figure 14(a)). The π-electron radical that is delocalized along the edges/defects exhibits fast spin–lattice relaxation through interaction with the adjacent π-electron system, which results in a broadening of the EPR signal. On the other hand, in the EPR spectrum of Pt/TiO_2-rGO-bR in the dark (Figure 14(a), black line), the broad signal disappears, while only the sharp signal is present. This is an indication that the bR protein predominantly adsorbs on the defect sites of rGO and "repairs" it.

Under illumination with light (λ > 440 nm), the Pt/TiO_2-rGO-bR nano-bio hybrid showed a decrease in signal corresponding to delocalized electrons in the rGO, which was accompanied by the appearance of signals that correspond to the localized electrons in anatase and rutile of titania P25 (Figure 14b). These data demonstrate that rGO upon excitation with visible light injects electrons into TiO_2. Upon turning off photoexcitation, the EPR signals of the TiO_2 lattice-trapped electrons were preserved at 4.5 K, indicating that the photogenerated charges are well

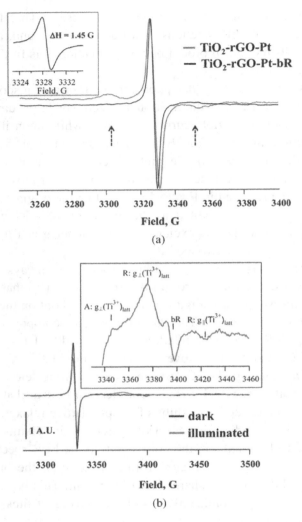

Fig. 14. (a) X-band EPR spectra of Pt/TiO$_2$-rGO (red line) and Pt/TiO$_2$-rGO-bR (black line) measured in the dark at 4.5 K; power = 2.08 mW, modulation amplitude = 8 G. Inset: EPR spectrum of Pt/TiO$_2$-rGO in the dark at 4.5 K; power = 2.08 mW, modulation amplitude = 1 G. (b) X-band EPR spectra of Pt/TiO$_2$-rGO-bR at 4.5 K in the dark (black line) and under illumination; λ > 440 nm (red line), power = 2.08 mW, modulation amplitude = 8 G. Inset: difference between spectra under illumination and dark, showing formation of signals that correspond to the lattice-trapped electrons in TiO$_2$: anatase (A) and rutile (R). [Reproduced with permission from ACS[59].]

separated, which results in their suppressed recombination. In this rGO-based photocatalyst bR also works as a light-harvesting entity capable of transferring electrons to TiO_2. Under white light illumination, there is probably combined input from rGO and bR.

To summarize, we have been successfully utilizing a natural proton pump bR and the PM architectures as a building module in the bio-assisted photocatalytic systems for the light-driven HER. In these nano-assemblies, bR serves as a visible-light harvester on Pt/TiO_2 photocatalyst and also contributes into optimizing of the protons-Pt catalyst interactions thus enhancing reduction of protons to hydrogen. Further introduction of 2-D carbon nanoscaffold rGO into the photocatalyst results in boosting performance of the nano-bio hybrid assembly under the visible light and remarkable reduction in the platinum co-catalyst content by 25%. As it was established by photoelectrochemical, EPR and TA experiment, rGO functions similarly to organic dye, a charge transfer can occur from rGO to TiO_2 NPs and, even more surprisingly, from rGO to bR. In addition, rGO provides a nano-scaffold for seamless interface between biological molecules, semiconductor particles, and platinum co-catalyst. One of the most attractive features of this approach is that all biological and inorganic materials can readily self-assemble without additional chemical coupling steps to form a stable and functional hierarchical photocatalytic system.

Conclusions and Outlook

Biological solar energy conversion provides an inspirational grounds as well as naturally-evolved architectures as practical materials for development of advanced hybrid photoenergy conversion systems. Driven by fundamental natural principles innovative scientific ideas have been conceived toward ultimate goal of advancing the field of artificial photosynthesis and, further, revolutionizing contemporary energy technologies. Ideally, prospective sustainable energy transformation system should be configures the way that it utilizes renewable sources (i.e. sunlight and salt water), includes inexpensive and earth-abundant, robust, yet environmentally-benign components (such as semiconductor metal oxide NPs) and efficiently produces clean zero-emission fuel (like hydrogen). Further, organic and natural dyes have been utilized for extending the visible light reactivity of semiconductor catalysts. However, hybrid systems which employ dyes or natural biomolecules isolated from original sources generally have limited stability under industry-relevant photocatalytic reactions conditions and relatively high cost to become commercially-feasible.

Here we described mechanisms of sunlight energy transformation via natural chlorophyll-independent pathway by a membrane proton pump machinery bacteriorhodopsin from salt loving microorganisms known as purple membranes. bR and it's membrane complex known as purple membranes are naturally evolved to tolerate and maintain their photoreactivity under demanding conditions, including exposure to light and oxygen, high ionic strength, temperatures, broad range of pH values. Next, we overviewed current successful examples of utilizing of the nature's "purple solar panels" in designed environmentally friendly bio-inspired photoenergy conversion schemes. Naturally-evolved structural and functional elegance of the natural purple "solar panels", their robustness and low cost can allow for overcoming limited stability, structural complexity and accessibility of other photocatalytic systems which include organic and biological structures. While all overviewed works represent convincing, yet still proof of a concept and there is much to be accomplished for translation of these concepts to practice. Further feasible steps can include but not limited to integration of biotechnological "farming" of Halobacteria with industrial processes such as saline chilling waters recycling, salt manufacturing from brines in natural or eco-friendly artificial pans, as well as using of components of waste and biomass compounds (e.g. alcohols, sugars, cellulose, lignin, EDTA) as sacrificial electron donors.

Acknowledgments

This work was performed at the Center for Nanoscale Materials, a U.S. Department of Energy, Office of Science, office of Basic Energy Sciences User Facility under Contract No. DE-AC02-06CH11357. Peng Wang thanks support by the National Basic Research Program of China (the 973 Program, No. 2013CB632401), the National Natural Science Foundation of China (NO. 21333006).

References

1. NRDC: Renewable Energy for America. http://www.nrdc.org/energy/renewables/solar.asp
2. T. Randall. Fossil Fuels Just Lost the Race Against Renewables. Bloomberg, April 14, 2015; Vol. 2015; http://www.bloomberg.com/news/articles/2015-04-14/fossil-fuels-just-lost-the-race-against-renewables
3. IEA: How solar energy could be the largest source of electricity by mid-century. Paris, 29 September 2014 Vol. 2015; http://www.iea.org/newsroomandevents/pressreleases/2014/september/how-solar-energy-could-be-the-largest-source-of-electricity-by-mid-century.html.
4. M. Roeb, M. Neises, N. Monnerie, C. Sattler and R. Pitz-Paal. Technologies and trends in solar power and fuels. *Energ. Environ. Sci.* **4**, 2011, 2503–2511.
5. G. Ma, T. Hisatomi and K. Domen. *Semiconductors for Photocatalytic and Photoelectrochemical Solar Water Splitting*, Springer International Publishing AG: Switzerland, 2015.
6. M. S. Dresselhaus and I. L. Thomas. Alternative energy technologies. *Nature* **414**, 2001, 332–337.
7. J. A. Turner. Sustainable hydrogen production. *Science* **305**, 2004, 972–974.
8. N. S. Lewis and D. G. Nocera. Powering the planet: Chemical challenges in solar energy utilization. *P. Natl. Acad. Sci. USA* **103**, 2006, 15729–15735.
9. M. F. Hohmann-Marriott and R. E. Blankenship. Evolution of Photosynthesis. *Annu. Rev. Plant. Biol.* **62**, 2011, 515–548.
10. A. Kay and M. Gratzel. Artificial Photosynthesis.1. Photosensitization of Tio2 Solar-Cells with Chlorophyll Derivatives and Related Natural Porphyrins. *J. Phys. Chem.-US* **97**, 1993, 6272–6277.
11. D. Gust, T. A. Moore and A. L. Moore. Mimicking photosynthetic solar energy transduction. *Accounts. Chem. Res.* **34**, 2001 40–48.
12. S. Y. Reece, J. A. Hamel, K. Sung, T. D. Jarvi, A. J. Esswein, J. J. H. Pijpers and D. G. Nocera. Wireless Solar Water Splitting Using Silicon-Based Semiconductors and Earth-Abundant Catalysts. *Science.* **334**, 2011, 645–648.
13. T. Hisatomi, J. Kubota and K. Domen. Recent advances in semiconductors for photocatalytic and photoelectrochemical water splitting. *Chem. Soc. Rev.* **43**, 2014, 7520–7535.
14. E. Reisner, J. C. Fontecilla-Camps and F. A. Armstrong. Catalytic electrochemistry of a [NiFeSe]-hydrogenase on TiO2 and demonstration of its suitability for visible-light driven H-2 production. *Chem. Commun.* 2009, 550–552.

15. D. K. Bora, E. A. Rozhkova, K. Schrantz, P. P. Wyss, A. Braun, T. Graule and E. C. Constable. Functionalization of Nanostructured Hematite Thin-Film Electrodes with the Light-Harvesting Membrane Protein C-Phycocyanin Yields an Enhanced Photocurrent. *Adv. Funct. Mater.* **22**, 2012, 490–502.
16. D. K. Bora, A. Braun and K. Gajda-Schrantz. *Solar Photoelectrochemical Water Splitting with Bioconjugate and Bio-Hybrid Electrodes*; Springer International Publishing: Switzerland, 2015.
17. A. Bachmeier, B. Siritanaratkul and F. A. Armstrong, *Enzymes as Exploratory Catalysts in Artificial Photosynthesis*; Springer International Publishing AG: Switzerland, 2015.
18. L. M. Utschig, S. R. Soltau and D. M. Tiede. Light-driven hydrogen production from Photosystem I-catalyst hybrids. *Curr. Opin. Chem. Biol.* **25**, 2015, 1–8.
19. D. K. Bora, A. Braun and E. C. Constable. "In rust we trust". Hematite — the prospective inorganic backbone for artificial photosynthesis. *Energ. Environ. Sci.* **6**, 2013, 407–425.
20. O. P. Ernst, D. T. Lodowski, M. Elstner, P. Hegemann, L. S. Brown and H. Kandori. Microbial and Animal Rhodopsins: Structures, Functions, and Molecular Mechanisms. *Chem. Rev.* **114**, 2014, 126–163.
21. D. S. W. Oesterhelt. Rhodopsin-like protein from the purple membrane of Halobacterium halobium. *Nature: New Biology* **233**, 1971, 149–152.
22. D. S. W. Oesterhelt. Functions of a New Photoreceptor Membrane. *Proc. Nat. Acad. Sci. USA* **70**, 1973, 2853–2857.
23. W. Kuhlbrandt, Bacteriorhodopsin — the movie. *Nature* **406**, 2000, 569–70.
24. I. R. Nabiev, R. G. Efremov and G. D. Chumanov. The Chromophore-Binding Site of Bacteriorhodopsin — Resonance Raman and Surface-Enhanced Resonance Raman-Spectroscopy and Quantum Chemical Study. *J. Bioscience.* **8**, 1985, 363–374.
25. R. Govindjee, S. P. Balashov and T. G. Ebrey. Quantum Efficiency of the Photochemical Cycle of Bacteriorhodopsin. *Biophys. J* **58**, 1990, 597–608.
26. Y. D. Jin, N. Friedman, M. Sheves, T. He and D. Cahen. Bacteriorhodopsin (bR) as an electronic conduction medium: Current transport through bR-containing monolayers. *P. Natl. Acad. Sci. USA* **103**, 2006, 8601–8606.
27. N. Hampp. Bacteriorhodopsin as a photochromic retinal protein for optical memories. *Chem. Rev.* **100**, 2000, 1755–1776.
28. R. Henderson. Purple Membrane from Halobacterium-Halobium. *Annu Rev Biophys. Bio.* **6**, 1977, 87–109.
29. D. Oesterhelt and W. Stoeckenius. Isolation of the cell membrane of Halobacterium halobium and its fractionation into red and purple membrane. *Methods in Enzymology.* **31**, 1974, 667–78.

30. A. Fujishima and K. Honda. Electrochemical photolysis of water at a semiconductor electrode. *Nature* **238**, 1972, 37–8.
31. X. B. Chen, L. Liu, P. Y. Yu and S. S. Mao. Increasing Solar Absorption for Photocatalysis with Black Hydrogenated Titanium Dioxide Nanocrystals. *Science.* **331**, 2011, 746–750.
32. X. B. Chen, S. H. Shen, L. J. Guo and S. S. Mao. Semiconductor-based Photocatalytic Hydrogen Generation. *Chem. Rev.* **110**, 2010, 6503–6570.
33. F. Gao, Y. Wang, D. Shi, J. Zhang, M. K. Wang, X. Y. Jing, R. Humphry-Baker, P. Wang, S. M. Zakeeruddin and M. Gratzel. Enhance the optical absorptivity of nanocrystalline TiO_2. film with high molar extinction coefficient ruthenium sensitizers for high performance dye-sensitized solar cells. *J. Am. Chem. Soc.* **130**, 2008, 10720–10728.
34. H. Tong, S. X. Ouyang, Y. P. Bi, N. Umezawa, M. Oshikiri and J. H. Ye. Nano-photocatalytic Materials: Possibilities and Challenges. *Adv. Mater.* **24**, 2012, 229–251.
35. K. Shimura and H. Yoshida. Heterogeneous photocatalytic hydrogen production from water and biomass derivatives. *Energ. Environ. Sci.* **4**, 2011, 2467–2481.
36. L. S. Li, T. Xu, Y. J. Zhang, J. Jin, T. J. Li, B. S. Zou and J. P. Wang. Photovoltaic characteristics of BR/p-silicon heterostructures using surface photovoltage spectroscopy. *J. Vac. Sci. Technol. A* **19**, 2001, 1037–1041.
37. R. M. N. Yerga, M. C. A. Galvan, F. del Valle, J. A. V. de la Mano and J. L. G. Fierro. Water Splitting on Semiconductor Catalysts under Visible-Light Irradiation. *Chemsuschem* **2**, 2009, 471–485.
38. K. Maeda. Photocatalytic water splitting using semiconductor particles: History and recent developments. *J. Photoch. Photobio. C* **12**, 2011, 237–268.
39. B. Oregan and M. Gratzel. A Low-Cost, High-Efficiency Solar-Cell Based on Dye-Sensitized Colloidal TiO_2 Films. *Nature* **353**, 1991, 737–740.
40. M. Gratzel. Photoelectrochemical cells. *Nature* **414**, 2001, 338–344.
41. M. Gratzel. Dye-sensitized solar cells. *J. Photoch. Photobio. C* **4**, 2003, 145–153.
42. N. K. Allam, C. W. Yen. R. D. Near and M. A. El-Sayed. Bacteriorhodopsin/TiO_2 nanotube arrays hybrid system for enhanced photoelectrochemical water splitting. *Energ. Environ. Sci.* **4**, 2011, 2909–2914.
43. N. Naseri, S. Janfaza and R. Irani. Visible light switchable bR/TiO_2 nanostructured photoanodes for bio-inspired solar energy conversion. *Rsc. Adv.* **5**, 2015, 18642–18646.
44. V. Thavasi, T. Lazarova, S. Filipek, M. Kolinski, E. Querol, A. Kumar, S. Ramakrishna, E. Padros and V. Renugopalakrishnan. Study on the Feasibility

of Bacteriorhodopsin as Bio-Photosensitizer in Excitonic Solar Cell: A First Report. *J. Nanosci. Nanotechno.* **9**, 2009, 1679–1687.
45. V. Renugopalakrishnan, B. Barbiellini, C. King, M. Molinari, K. Mochalov, A. Sukhanova, I. Nabiev, P. Fojan, H. L. Tuller, M. Chin, P. Somasundaran, E. Padros and S. Ramakrishna. Engineering a Robust Photovoltaic Device with Quantum Dots and Bacteriorhodopsin. *J. Phys. Chem. C* **118**, 2014, 16710–16717.
46. R. Mohammadpour and S. Janfaza. Efficient Nanostructured Biophotovoltaic Cell Based on Bacteriorhodopsin as Biophotosensitizer. *Acs. Sustain. Chem. Eng.* **3**, 2015, 809–813.
47. M. Gratzel. Recent Advances in Sensitized Mesoscopic Solar Cells. *Accounts Chem. Res.* **42**, 2009, 1788–1798.
48. A. Biesso, W. Qian, X. H. Huang and M. A. El-Sayed. Gold Nanoparticles Surface Plasmon Field Effects on the Proton Pump Process of the Bacteriorhodopsin Photosynthesis. *J. Am. Chem. Soc.* **131**, 2009, 2442–2443.
49. L. K. Chu, C. W. Yen and M. A. El-Sayed. Bacteriorhodopsin-based photoelectrochemical cell. *Biosens. Bioelectron* **26**, 2010, 620–626.
50. A. Biesso, W. Qian and M. A. El-Sayed. Gold nanoparticle plasmonic field effect on the primary step of the other photosynthetic system in nature, bacteriorhodopsin. *J. Am. Chem. Soc.* **130**, 2008, 3258–3259.
51. C. W. Yen, S. C. Hayden, E. C. Dreaden, P. Szymanski and M. A. El-Sayed. Tailoring Plasmonic and Electrostatic Field Effects To Maximize Solar Energy Conversion by Bacteriorhodopsin, the Other Natural Photosynthetic System. *Nano. Lett.* **11**, 2011, 3821–3826.
52. C. W. Yen, L. K. Chu and M. A. El-Sayed. Plasmonic Field Enhancement of the Bacteriorhodopsin Photocurrent during Its Proton Pump Photocycle. *J. Am. Chem. Soc.* **132**, 7250–+.
53. Z. L. Zhao, P. Wang, X. L. Xu, M. Sheves and Y. D. Jin. Bacteriorhodopsin/ Ag Nanoparticle-Based Hybrid Nano-Bio Electrocatalyst for Efficient and Robust H-2 Evolution from Water. *J. Am. Chem. Soc.* **137**, 2015, 2840–2843.
54. A. Rakovich, A. Sukhanova, N. Bouchonville, E. Lukashev, V. Oleinikov, M. Artemyev, V. Lesnyak, N. Gaponik, M. Molinari, M. Troyon, Y. P. Rakovich, J. F. Donegan and I. Nabiev. Resonance Energy Transfer Improves the Biological Function of Bacteriorhodopsin within a Hybrid Material Built from Purple Membranes and Semiconductor Quantum Dots. *Nano Lett.* **10**, 2010, 2640–2648.
55. Z. S. Lu, J. Wang, X. T. Xiang, R. Li, Y. Qiao and C. M. Li. Integration of bacteriorhodopsin with upconversion nanoparticles for NIR-triggered photoelectrical response. *Chem. Commun.* **51**, 2015, 6373–6376.

56. Z. B. Guo, D. W. Liang, S. Y. Rao and Y. Xiang. Heterogeneous bacteriorhodopsin/gold nanoparticle stacks as a photovoltaic system. *Nano Energy* **11**, 2015, 654–661.
57. S. Y. Rao, S. F. Lu, Z. B. Guo, Y. Li, D. L. Chen and Y. Xiang. A Light-Powered Bio-Capacitor with Nanochannel Modulation. *Adv. Mater.* **26**, 2014, 5846–5850.
58. S. Balasubramanian, P. Wang, R. D. Schaller, T. Rajh and E. A. Rozhkova, High-Performance Bioassisted Nanophotocatalyst for Hydrogen Production. *Nano Lett.* **13**, 2013, 3365–3371.
59. P. Wang, N. M. Dimitrijevic, A. Y. Chang, R. D. Schaller, Y. Z. Liu, T. Rajh and E. A. Rozhkova. Photoinduced Electron Transfer Pathways in Hydrogen-Evolving Reduced Graphene Oxide-Boosted Hybrid Nano-Bio Catalyst. *Acs. Nano* **8**, 2014, 7995–8002.
60. K. S. Novoselov, A. K. Geim, S. V. Morozov, D. Jiang, Y. Zhang, S. V. Dubonos, I. V. Grigorieva and A. A. Firsov. Electric field effect in atomically thin carbon films. *Science* **306**, 2004, 666–669.
61. Y. Saga, T. Watanabe, K. Koyama and T. Miyasaka, Mechanism of photocurrent generation from bacteriorhodopsin on gold electrodes. *J. Phys. Chem. B* **103**, 1999, 234–238.
62. S. A. Waschuk, A. G. Bezerra, L. Shi and L. S. Brown. Leptosphaeria rhodopsin: Bacteriorhodopsin-like proton pump from a eukaryote. *P. Natl. Acad. Sci. USA* **102**, 2005, 6879–6883.
63. C. Horn and C. Steinem. Photocurrents generated by bacteriorhodopsin adsorbed on nano-black lipid membranes. *Biophys. J* **89**, 2005, 1046–1054.

Chapter 8

Status and Perspectives on the Photocatalytic Reduction of CO_2

Jun Zhang and Zhaojie Wang

State Key Laboratory of Heavy Oil Processing
College of Chemical Engineering
China University of Petroleum
Qingdao 266580, PRC
zhangj@upc.edu.cn

1. The Background of CO_2 Scene

In the past decades, fossil fuels, such as coal, petroleum and natural gas have evolved into major conventional energy sources due to their availability, stability and high energy density.[1,2] Because of human activities, tremendous amount of these fuels are consumed each year. One of the most directly associated side results is the increased level of atmospheric CO_2, which has become the major contribution to the "greenhouse" effect. An estimated 37 Gt of CO_2 emissions per year are generated due to the heavy reliance on fossil fuels currently, and this number will rise to about 36–43 Gt by 2035, due to the policies made by governments and renewable energy sources.[3,4] Considering that the natural cycle of CO_2 emission and uptake (mainly fixed by plants, microorganisms and underground accumulation) involves a balance of ~90 Gt, it can be inferred that the anthropogenic emissions will disturb the balance remarkably.[4] According to the statistical analysis, CO_2 level of the atmosphere has reached 395 ppm in 2012 that exceeds the natural fluctuation of 180–300 ppm over the past 800,000 years.[5] The current highest value in the past 15 million years is considered to be a great factor for anthropogenic climate change such as the rise in global temperature, where a variety of overwhelming evidences have been illustrated.[6] As one of the greenhouse gases, CO_2 contributes to raising the global temperature through the absorption of infrared radiation from the sun. International Panel on Climate Change (IPCC) predicted that the atmospheric CO_2 level could reach up to 590 ppm by 2100 and the average global temperature will rise by 1.9°C. This increment is expected to result in the transformation of arable land to desert, melting of the ice caps and increasing precipitation across the globe, all of which will lead to loss of habitat and a rise in sea level.[7]

Apparently, the global atmospheric CO_2 level has already increased rapidly by 2.25 ppm per year. In other words, every year, more than 12 billion tons of CO_2 is released into the atmosphere.[8] However, as estimated by the U.S. Department of Energy, natural processes can only absorb about half of that amount. To mitigate the effect of increasing CO_2 emissions, plenty of researches have been focused on its treatment. Herein, there are three strategies available to reduce the amount of CO_2 in the atmosphere including direct reduction of CO_2 emission,[9] CO_2 capture and storage (CCS),[10,11] and CO_2 conversion and utilization.[12,13]

One of the effective approaches to reduce the CO_2 emission is the exploitation of new energy sources such as geothermal energy, wind energy, ocean energy, biomass energy, fusion energy and solar energy, etc. Although these energy sources can help reduce environmental pollution and decrease the dependence on the conventional fossil fuels, the diverse shortcomings such as low energy density and conversion efficiency, poor stability and limitation for continuous powering, large operating space and high cost are still the urgent issues to tackle. Meanwhile, the increasing utilization efficiency of fossil fuels to dramatically lower the CO_2 emission seems difficult due to the increasing population and demand in the world.

Besides mitigating atmospheric CO_2 emissions, effective techniques to rebalance the content of CO_2 and minimize greenhouse effect are thus more urgent and important than ever. Carbon capture and storage is a potential approach to alleviate this issue. CCS includes a portfolio of technologies, involving the core processes for CO_2 capture, separation, transport, storage and monitoring.[11] Usually, it refers to a number of technologies that capture CO_2 at certain stage in the corresponding processes such as combustion (most generally for power generation) or gasification. Many industrial processes, most notably cement manufacture, iron and steel making and natural gas treatment also intrinsically produce CO_2 and can be fitted with CO_2 capture technologies. By far, CCS has offered one of the very few remaining methods to reduce CO_2 emissions in these industries. The captured CO_2 is then pressurized to ~100 bar or more, prior to being transported to a storage site, where it is injected into one of numerous types of stable geological formation (geological reservoirs or ocean), trapping it for hundreds or thousands of years and preventing its subsequent emission into the atmosphere. During each link in the chain, the additional monitoring of CO_2 is of essence to ensure the normal operation. All of the individual components of the CCS chain, from capture all the way through to (and including) storage, have been demonstrated at or close to industrial scale. Many kinds of absorbent materials including nanoporous carbons, porous polymers, porous aromatic frameworks, metal organic frameworks (MOFs) and zeolite have emerged for CCS uses. An ideal absorbent material must possess the properties of high CO_2 adsorption capacity, excellent adsorption selectivity of CO_2 over other gases, good chemical and mechanical stability, large-scale synthesis with low cost and minimal regeneration energy demand, etc., which is still the limiting factor to apply CCS technology on a large scale. Furthermore, the individual step such as separation, purification, compression, transportation and storage requires the input of additional energy.

During the past decade, researchers have switched their attention to the efficient recycling of CO_2 and even artificial conversion of CO_2 to fuels or useful chemicals. The experimental investigations show that the strategy is feasible to deal with the excessive atmospheric CO_2. For instance, high energy compounds such as methanol or methane produced by CO_2 conversion, using solar energy are often referred to as solar fuels,[15,16] and then can be combusted or used in fuel cells. This closes the carbon cycle, since the external addition of CO_2 from fossil fuels is no longer in the cycle, with the term "methanol economy" suggested for such cycle.[17,18] Many other hydrocarbon fuels and valuable chemicals can also be produced from the atmospheric CO_2 by means of biological, thermochemical, electrochemical or photocatalytic methods.[19–21]

Biological method: Biomass to fuel production is a viable approach for carbon recycling.[22,23] The basic process involves biomass conversion via photosynthesis, conversion of biomass into fuels such as ethanol or biodiesel using biochemical processes, and use of the bioresidue for production of biogas (methane). Plenty of feedstock and algae or cyanobacteria systems have been applied in the biological conversion and the world fuel production from biomass has been increasing rapidly (i.e. 3.91 quadrillion Btu was produced from biomass in 2008, Department of Energy statistics). However, the sunlight-to-fuel energy conversion efficiency in photosynthesis is approximately 1%, and other energy inputs required for biomass processing lower the overall sunlight-to-biofuel conversion efficiency.[24] As a result, tremendous land/water areas will be required for biomass to meet the energy needs, even if, in a developing country. Furthermore, such massive cultivation of a single biological species will undoubtedly raise environmental challenges. Other issues such as the complexity, high cost and lack of robustness destine that biofuels will be a part of the future energy infrastructure but are unlikely to play a decisive role.[25–28]

Thermochemical method: The present methods for recycling CO_2 involve concentrating CO_2 from atmosphere or a point source, introducing it into typical chemical processes to produce value-added products, then applying these products and capturing the exhausted CO_2 to end the cycle. Although not yet fully developed, direct reduction of CO_2 via thermochemical cycles including one-step and multi-step thermochemical process was used to make synthetic gas. The following equation, which is endothermic (ΔG^0 of 257 kJ/mol) shows a single-step thermochemical decomposition process of CO_2 to carbon monoxide and oxygen,

$$CO_2 + \text{energy} \rightleftharpoons CO + 1/2\, O_2. \tag{1}$$

For a complete conversion of CO_2, a minimum temperature of ~3075°C is needed, at which ΔG^0 becomes zero.[29] However, in principle, the CO yield is only 30% near 2400°C, which is realistically much lower due to reverse reaction.[30] Alternatively, the reaction temperature can be significantly decreased and the product separation can be achieved with multi-step thermochemical cycles. Possibilities of using different multi-step thermochemical cycles for CO_2 conversion such as those based on cadmium, antimony, zinc, iron, and nickel were explored.[29] For example, the Zn/ZnO cycle is given by,

$$ZnO \rightleftharpoons Zn + 1/2 O_2 \quad \Delta H^0_{25°C} = +350.5 KJ/mol. \quad (2)$$

$$Zn + CO_2 \rightleftharpoons ZnO + CO \quad \Delta H^0_{25°C} = -67.5 KJ/mol. \quad (3)$$

Reaction 2 is highly endothermic and requires a temperature of ~1700°C to dissociate the metal oxide into gas phase zinc and oxygen (the efficiency estimated is about 39% under concentrated solar radiation). The produced gases can either be quenched or separated after reaction 2. Reaction 3 to produce CO is exothermic, and thus an operating temperature between 425°C and 725°C is suggested. Despite these isolated efforts and excellent ideas, thermochemical approaches of CO_2 reduction associated with high temperature operation in a reactive environment have yet proved their viability.[31,32]

Electrochemical method: Following the concept of water electrolysis, a number of studies have explored the use of electrochemical catalysts to reduce CO_2 dissolved in liquids. A wide range of products, even C6 hydrocarbons including paraffins and olefins, with high energy density can be directly synthesized.[33] Generally, the process involves electrolysis of a solution containing dissolved CO_2 by applying a voltage greater than that necessitated by thermodynamics between two immersed electrodes. The standard redox voltage per electron for CO_2 splitting is 1.33 V vs. at 25°C and 1 atm. If the electrode overpotential and ohmic losses are added to thermoneutral voltage (~1.47 V, similar to that for water electrolysis), a voltage of about 2 V or more would be generally required to drive the reaction in aqueous electrolytes.[34] In an electrochemical splitting system, electrode materials and the potential applied to the electrodes play major roles in determining the products, selectivity and kinetics, while CO_2 concentration in the electrolyte and mass transport are decisive factors in determining the formation efficiencies of the different products.[35] Besides, the solubility of CO_2 in aqueous solution is as low as 33 mM at 25°C at 1 atm. In spite of the moderate operating temperature, CO_2 pressure and even other factors to enhance the solubility, novel electrode different from the traditional ones should also been found to improve mass transport and hence product formation rate and selectivity.

Photocatalytic method: It should also be noted that, apart from the various problems as referred, the above approaches are not implemented in large scale and difficult to provide bulk chemicals and/or fuels via simple treatment. That is, CO_2 must be chemically reduced, which requires a substantial input of energy. From a sustainable viewpoint, sunlight is the ideal energy source. The interest in such field has been aroused dramatically after 1970s, when photoelectrochemical reduction of CO_2 to organic compounds was demonstrated.[36] Direct solar conversion of CO_2 into hydrocarbon fuels using sunlight is an attractive prospect to reduce atmospheric CO_2 concentrations while providing a renewable energy dense portable fuel compatible with our current energy infrastructure. As expected in the long term usage, artificial photosynthesis, photocatalytic conversion of CO_2 in combination with photocatalytic H_2O splitting using solar energy is the most attractive route.[37]

Since CO_2 is a stable molecule, besides the energy demand, the application of extremely talented catalysts which is capable of driving its selective conversion into target chemicals is required as well. The activity and stability of a catalyst at normal operating conditions represent its acceptability to researchers. In fact, the state of the art in the photocatalytic CO_2 reduction is far from being optimal and there are still considerable breakthroughs to be made before it can be considered as a viable economical process.[37] For example, the reduction of CO_2 and specifically the formation of CH_3OH and CH_4 require a hydrogen source. Among the various possibilities, the most attractive but scientifically more challenging is the use of water splitting hydrogen as the electron donor. The process of the photocatalytic reduction of CO_2 by H_2O is considerably more difficult than the overall water splitting, which puts forward the higher request to the multifunction photocatalysts.[38] Another remarkable challenge is that the vast majority of photocatalysts that have some efficiency for CO_2 photoreduction do not exhibit photoresponse to visible-light; and therefore it is a request to develop visible-light driven photocatalysts for this process.[37,39,40] Challenges and developments in this area, specially related to novel catalytic materials with excellent efficiency, high yield and selectivity, still leave a lot to be desired to the recycle of CO_2 to carbonaceous fuels under solar irradiation without using extra energy.

2. Basic Principles of Photocatalytic Conversion

The characteristic of a photocatalytic reaction involves active electrons and holes, which endow the requirement of special electronic states of the aimed catalysts. Fortunately, the developed semiconductors are attractive type of photocatalyst materials due to their energy band.[41] The valence band (VB) occupied by electrons in the ground state owns the energy state below the gap, whereas the states above the gap form the unoccupied conduction band (CB). These two outermost energy levels are responsible for solid-state properties, conductivity, and reactivity according to the classical band theory of solids.

At a certain high temperature, some electrons in VB can be thermally excited to the CB and the resulting changed electron-density distribution can be characterized by the Fermi level of the semiconductor. According to the bandgap model for photocatalysis proposed by Demeestere *et al.* after absorption of a photon of energy equal to or higher than the bandgap, an electron can also be excited from the VB into the CB, thereby leaving an empty state that constitutes a quasiparticle referred to as a hole.[42] The sum of both charge carriers, that is, electrons and holes, in an illuminated semiconductor are larger than in the genteel due to the creation of electron–hole pairs, as shown in Fig. 1. The carriers, once separated in a certain space, will experience the intraband transitions by the different pathways in the semiconductor. The most useful migration is to the surface of the photocatalyst and then the separated carriers get trapped at the trap sites. Eventually the electron–hole pairs transfer to the absorbed acceptor molecules surrounded the semiconductor, thereby initiating the corresponding reduction or oxidation reactions.[43]

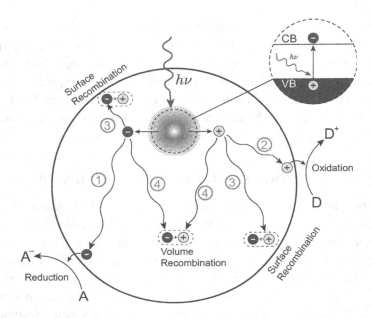

Fig. 1. Photoinduced formation of an electron–hole pair in a semiconductor with possible decay paths. A=electron acceptor, D=electron donor. Adapted from Ref. 43 with permission.

The tendency of the molecular orbital of the adsorbed species to accept/donate electrons manifests itself in its reduction/oxidation potential, respectively. Thermodynamically, the ability of a semiconductor to transfer photogenerated electrons and holes to adsorbed molecules depends on the alignment of the quasi-Fermi levels of the carriers with the redox potential.[44,45] As is shown in Fig. 2, the reaction conditions are determined by the position of the band edges. Generally, the CB edge must lie above the lowest unoccupied molecular orbital (LUMO) of the acceptor molecule due to the need of reduction processes. That is to say, since the level of the redox potential goes from positive to negative with increasing energy, the potential level of the CB has to be more negative than the reduction potential of the absorbed acceptor. In contrast, the transfer of a photogenerated hole from the semiconductor to an adsorbed molecule, equivalent to the transfer of an electron in the opposite direction, requires the energy level of the highest occupied molecular orbital

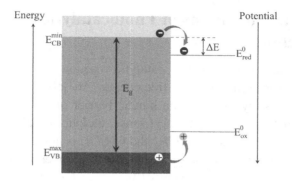

Fig. 2. Thermodynamic constraints on the transfer of charge carriers to the adsorbed molecules.

(HOMO) of the oxidized molecule to lie above the VB or — in terms of redox potentials — it has to be more negative than the potential of the valence band.[39,46]

The redox reaction sites may be located either directly on the surface of the semiconductor close to the photoexcitation or indirectly across the interface at the homo or heterojunction in the composite catalyst.[47] Therefore, the primary role of the semiconductor in photocatalysis is absorbing an incident photon, generating an electron–hole pair, and facilitating its separation and transport, while catalysis of the reaction is an additional function which is usually performed by a different material.

However, unsatisfying amount of excited electron–hole pairs can successfully transfer to participate the redox reaction. Recombination of photoexcited electrons and holes prevents them from transferring to the surface to react with absorbed species. Lifetime of the electron–hole pairs is only a few nanoseconds, which is adequate for promoting redox reactions. In comparison, the time scale of electron–hole recombination is two or three orders of magnitude faster than other electron transfer processes.[37] This is thus considered as one of the major limiting steps in photocatalysis. Therefore, any process which inhibits electron–hole recombination would greatly increase the efficiency and improve the rates of CO_2 photoreduction. The details of engineering bandgap for light harvesting and electron transport are discussed in Sec. 6.

3. General Conditions of Photocatalytic CO_2 Reduction

From a kinetic point of view, a catalytic reaction is characterized by a negative difference in the Gibbs free energy ($\Delta G^0 < 0$), in which the catalyst is used feasibly to lower the kinetic barrier of the reaction without changes on itself.[48,49] In the strictest sense, the photocatalytic CO_2 reduction into hydrocarbons such as CH_4 and CH_3OH is not a catalytic process, because of the uphill reaction. It means that the conversion brings about a positive change in Gibbs free energy ($\Delta G^0 > 0$), which determines the extra input energy by incident light being involved.

$$CO_2 + 2H_2O \rightarrow CH_3OH + 3/2O_2 \quad \Delta G^0 = 702.2 \text{ kJ/mol}$$
$$CO_2 + 2H_2O \rightarrow CH_4 + 2O_2 \quad \Delta G^0 = 818.3 \text{ kJ/mol}$$

In fact, it represents an example of artificial photosynthesis precisely.[50–52] Nonetheless, in the usually accepted sense, we still follow the broader definition of photocatalysis throughout this book to comply with the vast majority of publications on the subject.

CO_2, a linear molecule, is one of the most thermodynamically and kinetically stable compounds of carbon. Photocatalytic reduction of CO_2 into hydrocarbon fuels, demands input energy to break $C=O$ bond and form $C-H$ bond, involving the participation of multiple electrons and a corresponding number of protons.[53] CO_2 can only be reduced with the equality of reducing agents for the reason that it keeps the highest chemical state (C^{4+}) of C atoms in CO_2. Obviously, the reduction half-reaction in the photocatalytic redox system only consumes the photogenerated electrons. In an ideal case, H_2O is a preferred candidate of reducing agents compared with other species such as H_2, S^{2-}, SO_3^{2-} and amines, owing to its richness, non-toxicity, and effectiveness. On the other hand, the oxidation half-reaction will lead to the generation of oxygen or hydrogen peroxide through the oxidation of water. However, very few systems have been shown to simultaneously reduce CO_2 and oxidize water, as the later process is difficult to accomplish.[54,55] Therefore, the focus of the research on CO_2 conversion remains on the reduction, whilst the oxidation is achieved by a surrogate reaction involving sacrificial electron donors.

Even though the researches on the photocatalytic CO_2 reduction have developed for nearly half a century, there is a long way to go in terms of efficient and commercially viable devices. As far as the result is concerned, the highest rates of aimed product formation generally do not exceed tens of μmol of product, e.g. methane, per hour of irradiation per gram of photocatalyst. Therefore, the urgent requirement in the field of CO_2 recycling technology is to develop visible-light-sensitive photocatalyst with high efficiency. Different sorts of photocatalysts, such as titanium and non-titanium containing solids, metal sulfides, phosphides, MOFs, organic molecules and various combinations, have been already introduced by many researchers. While some of the catalysts exhibited high conversion rates and selectivity under visible-light irradiation, other catalysts were not feasible for reaction with high yield under visible-light. Great efforts have been made by researchers to improve the properties of catalysts in terms of CO_2 conversion. The consequences of these catalysts, along with possible approaches to design them rationally, are discussed in detail in Secs. 4 and 5.

Chemical Pathways: The transfer of an electron to the adsorbed molecule initiates different possible chemical reactions which determine the final outcome and efficiency of the phtocatalytic process. In the case of CO_2 reduction, the multi-step reaction contains up to twelve electrons and protons transfer (the breakage of C–O bonds and the formation of C–H bonds). Of course, different courses result in the formation of different products, i.e. $CO_2 \rightarrow HCOOH \rightarrow HCHO \rightarrow CH_3OH \rightarrow CH_4$ or $CO_2 \rightarrow CO \rightarrow C\bullet \rightarrow CH_2 \rightarrow CH_4$, to yield the end product CH_4 through different intermediates.[56–59] According to the published literatures, the exact mechanism of photocatalytic CO_2 reduction still remains unraveled. Nonetheless, more information is continuously being gathered based on the numerous experimental methods (gas and ion-exchange chromatography,[60–62] mass spectrometry, IR,[63–65] EPR,[66–68] Auger electron and X-ray photoelectron spectroscopy,[69] transient absorption measurements, scanning tunneling microscopy,[70] and many surface-oriented studies[49]) and computational methods. The experimental techniques are applied mostly to analyze semiconductor excitation, charge-carrier lifetimes, mobility and separation, while the computational methods provide critical information regarding the adsorption of the reagents on the photocatalyst surface as well as the structure and energetics of the putative intermediate products and transition states.

Photocatalytic reduction of CO_2 by using water as a reductant can be assumed as follows (Eqs. (5–11)).[71] The photo excited electrons and holes in the lattice are separated and trapped by appropriate sites of catalyst to avoid recombination. The holes oxidize water on the surface (Eqs. (1–3) and provide protons for the reduction.[72]

Photooxidation reactions

Water oxidation/ decomposition	$2H_2O + 4h^+ \rightarrow O_2 + 4H^+$	(1)	$E^0_{redox} = 0.82$ V
Hydrogen peroxide formation	$2H_2O + 2h^+ \rightarrow H_2O_2 + 2H^+$	(2)	$E^0_{redox} = 1.35$ V
Hydroxyl radical formation	$H_2O + h^+ \rightarrow OH^{\bullet} + H^+$	(3)	$E^0_{redox} = 2.32$ V

Photoreduction reactions

Hydrogen formation	$2H^+ + e^- \rightarrow H_2$	(4)	$E^0_{redox} = -0.41$ V
CO_2 radical formation	$CO_2 + e^- \rightarrow CO_2^{\bullet-}$	(5)	$E^0_{redox} = -1.90$ V
Formic acid formation	$CO_2 + 2H^+ + 2e^- \rightarrow HCOOH$	(6)	$E^0_{redox} = -0.61$ V
Carbon monoxide formation	$CO_2 + 2H^+ + 2e^- \rightarrow CO + H_2O$	(7)	$E^0_{redox} = -0.53$ V
Formaldehyde formation	$CO_2 + 4H^+ + 4e^- \rightarrow HCHO + H_2O$	(8)	$E^0_{redox} = -0.48$ V
Methanol formation	$CO_2 + 6H^+ + 6e^- \rightarrow CH_3OH + H_2O$	(9)	$E^0_{redox} = -0.38$ V
Methane formation	$CO_2 + 8H^+ + 8e^- \rightarrow CH_4 + 2H_2O$	(10)	$E^0_{redox} = -0.24$ V
Ethanol formation	$2CO_2 + 12H^+ + 12e^- \rightarrow C_2H_5OH + 3H_2O$	(11)	$E^0_{redox} = -0.16$ V

By comparison, the single electron reduction of CO_2 to an anion radical $CO_2^{\bullet-}$ (Eq. (5)) has a rather negative electrochemical potential of -1.9 V vs. the normal hydrogen electrode (NHE), which is much more negative than

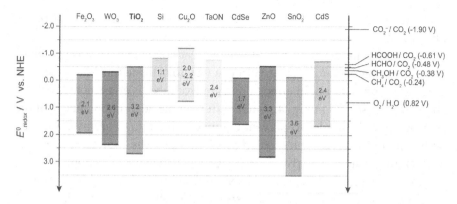

Fig. 3. Conduction band, valence band potentials, and bandgap energies of various semiconductor photocatalysts relative to the redox potentials at pH 7 of compounds involved in CO_2 reduction. Adapted from Ref. 48.

conduction band of most semiconductors (Fig. 3). Although this step seems difficult, the situation is better for a proton-assisted transfer of multiple electrons. Eqs. (6–11) list the corresponding electrochemical CO_2 reduction to formic acid, carbon monoxide, formaldehyde, methanol, methane and ethanol, at pH = 7, respectively. The redox potentials are comparable to proton reduction itself (Eq. (4)) of a one-electron process. And the superiority is the potentials are even less negative than the CB of many typical semiconductors, which make these reactions feasible.

From a thermodynamic point of view, the formation of methane and methanol are more favorable in CO_2 reduction, since these reactions take place at lower potentials. However, in the view of kinetics, the drawback makes methane and methanol formation more difficult than carbon monoxide, formaldehyde and formic acid because more electrons are required for the former reactions. Due to the complicated nature of the inorganic photocatalyst surfaces, the interaction between photocatalysts and absorbed species may undergo a series of one-electron transfer processes instead of a multi-electron, multi-proton transfer process. Apart from the kinetic difficulty, there is little evidence in the literature of multi-electron transfer processes.[47] With this concern, the reaction is likely to be initiated by single electron reduction of CO_2 to $CO_2^{\cdot-}$ and experiences a series of one-electron steps. Combined with the two points of view, the first electron transfer with the potential around −1.9 V vs. NHE remains a severe obstacle to the photoreduction of CO_2, and likely constitutes a strongly limiting step.

An important element to facilitate the reaction is that the adsorption of CO_2 on a semiconductor surface offers a way to activate the otherwise inert molecule for reduction. In effect, there are five models constructed for CO_2 absorption as shown in Fig. 4, which differs in the adsorption energy of the system.[73] Here, in order to prevent confusion, the two O atoms connected with centered C atom were named as O_a and O_b. The first cartoon is that the CO_2 molecule is linearly adsorbed on the surface with only one O_a atom (Fig. 4(a)). If the solid surface absorbed C atom rather than O atoms, it generates a monodentate carbonate species (Fig. 4(b)). In the third, both the O_a and C atom of CO_2 molecule are connected with two metal atoms to form a bidentate carbonate species (Fig. 4(c)). The fourth is the generation of a bridged carbonate geometry with the C atom of CO_2 pointing downward and two O atoms of CO_2 binding with two metal atoms to form a C···O bond with the O atom on the surface (Fig. 4(d)). Different from the fourth, the bridging C atom is points upwards instead of downwards towards the two metal atoms. (Fig. 4(e)). The formation of the last two models relies on the presence of an M-O-M bond on the surface.

The activation of chemically stable CO_2 with a closed-shell electronic configuration, linear geometry, and $D_{\infty h}$ symmetry initiates multistep reactions most likely featuring one electron transfer to CO_2 to generate $CO_2^{\cdot -}$ species.[37] Because of the introduction of a single electron, a bending of the molecular structure may take place between the newly acquired electron situated on the electrophilic carbon atom and the free electron pairs on the oxygen atoms. It will cause the high energy of the LUMO of CO_2 and very low electron affinity of the molecule due to the loss of symmetry and the

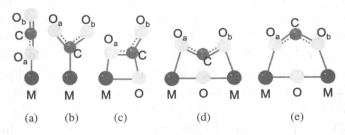

Fig. 4. Possible configurations of adsorbed CO_2 on the photocatalyst surfaces.

increased repulsion between these free electron pairs in the bent structure with a C_{2v} symmetry.[74] The nonlinear CO_2 molecule generated on the surface of photocatalyst is more destabilized than the linear one, exhibiting the high reactivity for photoreduction.[75] Once the surface-bound $CO_2^{\cdot-}$ formed, the reduction proceeds a series of elementary steps including transfer of electrons, protons and hydrogen radicals, as well as breakage of C–O bonds and formation of new C–H bonds.

What's interesting is that the different binding mode of $CO_2^{\cdot-}$ on the catalyst surface starts the different pathways. For example, the monodentate binding via one of the O atoms to a heterogeneous atom (Fig. 4(a)) or via the C atom to the surface generally favors the formation of the carboxyl (hydroxyformyl) radical •COOH.[75] While, bidentate binding of the CO_2 through both oxygen atoms to two surface atoms results in the preferential attachment of a hydrogen atom to the carbon atom, which leads to the formation of a formate anion bound in a bidentate mode.[72] The whole reaction process is more complex out of sight. Several intermediates are radical species, whose recombination at different stages partially accounts for the number of possible pathways leading to different final products. Taking methane as an example, there are three full pathways proposed in the literature for the conversion of CO_2 into methane, which can be referred to as (1) the formaldehyde pathway, (2) the carbene pathway, and (3) the glyoxal pathway, on account of their unique intermediate in the following (Fig. 5).[76] All these exact order and mechanistic details of each subsequent step remains to be validated.

4. Rational Design of TiO_2 Photocatalyst

4.1 State-of-the-art TiO_2 photocatalyst

In the majority of the photocatalytic reactions for environmental remediation and pollutant degradation, TiO_2 is the best performing semiconductor. The advantages of titania photocatalysts, such as strong resistance to chemical and photocorrosion, low operational temperature, low cost, significantly low energy consumption, have led to the relevant applications of the photocatalytic reduction of CO_2. However, TiO_2 exhibits a relatively high energy bandgap (3.2 eV) and can only be excited by high energy UV irradiation with a wavelength shorter than 387.5 nm.

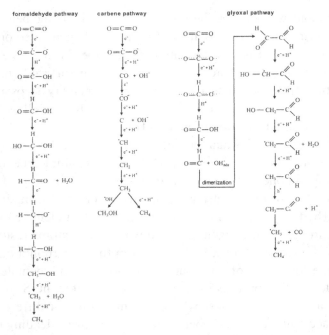

Fig. 5. Three proposed mechanisms for the reduction of CO_2 to methane: formaldehyde, carbene and glyoxal pathways.

As shown in the early work by Fujishima and Honda *et al.* the photocatalytic splitting of water and reduction of CO_2 using TiO_2 is applicable.[56,77] With its applications as a photoanode in inorganic and hybrid solar cells[78–80] as well as in photo- and electrochromic devices,[81] TiO_2 has attracted enormous interest in terms of studies on its crystal lattice, surface, charge transport, and other relevant properties.[82,83]

Particularly worth mentioning, P-25 from Evonik (formerly Degussa) is a standard material in the field of photocatalysis. It's a TiO_2 powder composed of about 80% anatase and 20% rutile, whose particle size usually ranges from 20 to 80 nm in diameter (average in ~25 nm). The effective surface area (BET) is known to be 49~56 m^2g^{-1}. Plenty of researches have been reported based on P-25[84,85] and these results provided the benchmark system in experiments focused on novel materials and other systems, because the preparation technique was relatively robust and the product was the most typical and commercial TiO_2 with reasonable performance.

At present, the general accepted view on the high activity of P-25 is attributed to complementary effects of the two constituting TiO_2 phases.[86–88]

It has been concluded that a problem in employing TiO_2 as the photocatalyst for CO_2 reduction irradiated with UV light is the lack of selectivity in the product distribution. Thus, although presumably CO_2 conversion using TiO_2 under UV light should be among the highest of all semiconductors, the product distribution is complex and covers almost all the products mentioned above and therefore the applicability of this process, even considering the use of UV light, is very limited. In addition to higher activity (ideally under visible-light irradiation) it is necessary to control the selectivity of the photocatalytic reaction to have some chance of applicability for the photocatalytic production of hydrocarbons.

The nature of the solvent plays a role in the product selectivity during photoreduction of CO_2 using TiO_2. In this regard, Kaneco and co-workers reported photocatalytic reduction of CO_2 using TiO_2 powders in liquid CO_2 medium working with light of wavelengths above 340 nm.[89] The addition of H_2O as the hydrogen donor was performed at the end of the irradiation time. The influence of temperature and pressure was also studied for 30 h illumination using TiO_2 powders, and it was found that the yield of HCOOH was almost constant and not affected either by temperature (0 to 25°C) or pressure (49.3 to 79.0 atm). Interestingly, no gaseous photoreduced products were observed under these conditions. The production of HCOOH almost linearly increased with the amount of TiO_2 in the photoreactor. The influence of different hydrogen donors namely CH_3OH, ethanol, isopropanol, propan-2-ol, nitric acid, hydrochloric and phosphoric acid, added at the end of the irradiation, was also studied in order to optimize the products. For example, the addition of phosphoric acid in the supercritical fluid CO_2 increased the yield of HCOOH. The beneficial influence of phosphoric acid can be understood considering that the protons decrease the potential of CO_2 reduction.

4.2 Rational design of TiO_2 catalyst

Efforts on two general aspects have been attempted to break through the efficiency, for examples, rational design of TiO_2 catalyst, and optimization of reaction conditions such as reductants, temperature, pressure, light intensity and wavelength etc. The efficiency for the photocatalytic reduction of CO_2 to produce desirable fuels strongly depends upon the type of the employed TiO_2 photocatalyst.

Small particle TiO_2 catalyst: One important factor to influence the efficiency of TiO_2 catalyst is the size. Taking the nanoparticle as an example, decreasing the particle size in nanoscale makes the bandgap of TiO_2 semiconductor larger and the surface area increased correspondingly. The external specific surface area determines available active sites for the surface reaction. Higher yields of methanol and methane over the TiO_2 nanoparticles are certified as the particle size reduced from 29 nm to 14 nm (Fig. 6(a)) and anatase nanoparticles with a diameter of 14 nm gave the highest yields.[90] At this situation, the decreased performance with larger particles may be due to smaller surface areas and longer electron migration paths. However, when the particle size is further reduced to 4.5 nm, CO_2 reduction rate is dropped associated with the changes in optical and electronic properties of the nanometer crystals especially with a widening of the bandgap and associated blue shift of the onset of absorption.[91–93] According to theoretical models, particle size of TiO_2 is an essential parameter for its photoactivity as well. The competing effects of effective particle size on light absorption and scattering efficiency, charge-carrier

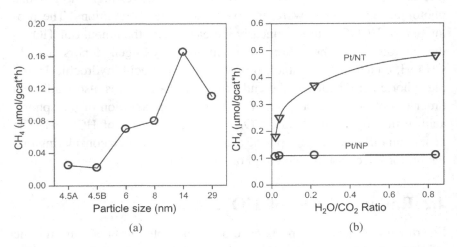

Fig. 6. Photocatalytic properties of TiO_2 for CO_2 conversion, (a) particle size, (b) nanoparticle vs. nanotube.

dynamics, and surface area leads to result that the optimal particle size might exist between 25 nm and 40 nm.[94]

Nanostructured TiO_2 photocatalyst: Besides change in particle size, another physical approach to enhance the photocatalytic properties of TiO_2 is the use of nanostructured materials. Advanced and various TiO_2 nanoarchitectures have been engineered for the purpose. Among these nanostructures, one-dimensional (1D) TiO_2 nanomaterials, such as nanorods, nanotubes and nanoribbons, are considered to own higher photocatalytic activity than nanoparticles and films. In these materials it appears that electron transport vectorially along the long axes of the crystallite is very fast and charge separation is favored due to the high electron mobility.[95,96]

TiO_2 nanotubes and nanorods can be formed in various ways. As reported the simplest method for the synthesis of nanorods is digesting TiO_2 nanoparticles at high temperatures with the presence of strong alkali. Annealed under solvothermal conditions, nanotubes can be obtained from layered titanates as starting precursors. Nanorods and nanotubes often exhibit enhanced photocatalytic activity with respect to nanoparticles. For instance, the yield difference of methane formation between TiO_2 nanotubes and nanoparticles has been investigated as the ratio of reactants $H_2O:CO_2$ equals 0.8.[95] The conversion rate to CH_4 is 0.48 $\mu mol/(g_{cat}\ h)$ for Pt(0.15 wt.%)/TiO_2 nanotube arrays which is nearly five times over that of Pt(0.12 wt.%)/TiO_2 nanoparticles [0.10 $\mu mol/(g_{cat}\ h)$, Fig. 6(b)]. However, several reports argue that nanoparticles exhibit higher performance for CO_2 reduction compared to other counter parts.[97] It should, however, be noted that there are considerable difficulties to make a rigorous analysis including further characterization of the product as well as examining composition, material size and loading effects to ensure this trend.

It is worth to mention that there is one further development of the use of TiO_2-nanotube or nanorod morphology which is the alignment at macroscopic distances of these nano objects. Therefore, not only by scattering but also through so-called "crystal photonic" effect (appears when the wavelength of the photon matches the distances between the TiO_2 rods and particles) can the TiO_2 nanotubes trap light.

Different crystal types: TiO_2 exists mainly in four different polymorphs (crystallographic phases): rutile (tetragonal), anatase (tetragonal), brookite (orthorhombic), and $TiO_2(B)$ (monoclinic).[82] However, from the viewpoint of photocatalysis, rutile and anatase phases indicate better efficiency under light radiation. Both of the structures can be characterized as chains of elongated TiO_6 octahedra with each Ti^{4+} ion surrounded by six O^{2-} ions.[43,83] The difference of the two configurations in structural symmetry and distance between atoms is shown in Fig. 7. The differences in the crystal structures are responsible for the different bandgaps of anatase (3.2 eV) and rutile (3.0 eV), which is mainly due to the position of the CB edge lying 0.2 eV higher for anatase.[98] The minimum energy of photons required to initiate an interband transition lies, therefore, in the UV range, with 389 nm for anatase and 413 nm for rutile. The only indirect interband transitions allowed by the crystal symmetry indicate that when an electron is photoexcited it requires the photon providing the energy and change in momentum. This decreases the photonic efficiency, as the rate of absorption events that yield available charge carriers is limited. Additionally, the rate of recombination of the photogenerated charge-carrier pairs can reach 90% within a period of 10 ns after their generation, thereby significantly limiting the number of charge carriers available for surface reactions.[82,99]

The studies on pure phase of TiO_2 show that anatase phase is more active than rutile, possibly attributed to the combined effect of lower recombination rate of electron–hole pairs and higher surface adsorptive capacity.[100–102] The interesting synergistic effect between anatase and rutile was observed, in which the addition of rutile significantly enhances the activity of anatase.[103–106] Based on the relative CB positions of anatase and

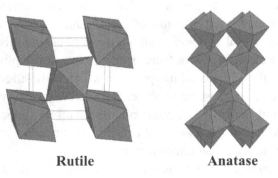

Fig. 7. Representations of the TiO_2 anatase and rutile forms.

rutile, Bickley and co-workers proposed a model where the electron transfer from photoactivated anatase to rutile (Fig. 8(a)).[107] Gray et al. studied mixed-phase TiO$_2$ materials using low temperature electron paramagnetic resonance (EPR) spectroscopy.[108] Trapped electrons in anatase were observed under visible-light irradiation which can only activate the rutile phase in the mixed-phase composite. A mechanism for the synergistic effect was proposed to involve electron transfer from the rutile CB to anatase trapping sites (Fig. 8(b)).

Nair et al. proposed an "interfacial model" (Fig. 9) to explain the synergistic effect between anatase and rutile.[109] The "interfacial model" considers band bending at the interfacial region between anatase and rutile. The band structure of the interfacial region in the dark (Fig. 9(a)) is almost the same as under UV irradiation (Fig. 9(b)), although the Fermi level shows an upward shift when both anatase and rutile are activated (Fig. 9(b)). According to the "interfacial model," electron transfer occurs from photoactivated anatase to rutile under UV irradiation since anatase

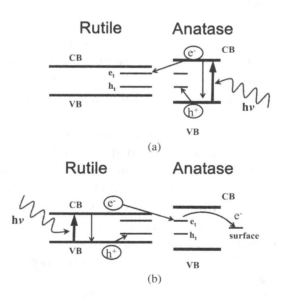

Fig. 8. Models for spatial charge separation in mixed-phase TiO$_2$: (a) Bickley and co-workers' model showing charge separation in anatase and electron trapping in rutile; (b) Gray, Rajh, and co-workers' model of a rutile antenna and subsequent charge separation.

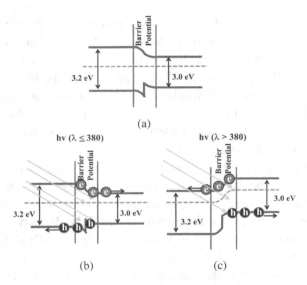

Fig. 9. A proposed interfacial band model (a) in the dark, (b) under UV irradiation, and (c) under visible-light. The anatase phase (bandgap 3.2 eV) is always shown on the left side and the rutile phase (bandgap 3.0 eV) is shown on the right.

has a more negative CB energy. However, under light irradiation with wavelength greater than 380 nm only the rutile phase is activated and the rutile CB shifts upward upon accumulation of photoexcited electrons, making it possible for the photoexcited electrons in rutile to reach the anatase CB (Fig. 9(c)).

Single crystal TiO_2 with exposed active facets: With a well-defined catalyst surface such as a single crystal, detailed information on the reaction mechanism can be obtained at the molecular level. From this point of view, the photocatalytic CO_2 reduction with H_2O on rutile-type TiO_2 (100) (Ti terminating) and (110) (O terminating) single crystal orientation surfaces have been performed (Fig. 10).[75,110] As shown in Table 1, the efficiency and selectivity of the photocatalytic reactions strongly depend on the exposed facet of TiO_2 single crystal. UV-irradiation of the TiO_2(100) single crystal catalyst in the presence of a mixture of CO_2 and H_2O led to the evolution of both CH_4 and CH_3OH at 275 K. Both of them are produced in the gas phase at the same time over a TiO_2 (100) surface, whereas only CH_3OH was detected with the TiO_2 (110) single crystal catalyst.

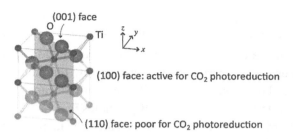

Fig. 10. Comparison of the (1 0 0) and (1 1 0) crystal faces in the crystal structure of rutile TiO_2 and their performance in photocatalytic CO_2 reduction.

Table 1. Yields of the formation of CH_4 and CH_3OH in the photocatalytic reduction of CO_2 (124 μmolg^{-1}) with H_2O (372 μmolg^{-1}) at 275 K.

Single crystal	Yield of CH_4 (μmolh^{-1}g^{-1})	Yield of CH_3OH (μmolh^{-1}g^{-1})
TiO_2 (100)	3.5	2.4
TiO_2 (110)	0	0.8

It is likely that the surface Ti atoms act as a reductive site. According to the surface geometric models for TiO_2(100) and TiO_2(110), the atomic ratio (Ti/O) of the top-surface Ti and O atoms which have geometric spaces large enough to have direct contact with CO_2 and H_2O molecules, is higher on TiO_2(100) than on TiO_2(110) surface. In the excited state, the surface with a higher Ti/O surface ratio, i.e. TiO_2(100), exhibits a more reductive tendency than (110). Such a reductive surface allows reduction of CO_2 to CH_4 molecules. In the HREELS spectrum of a clean TiO_2(100) single crystal surface after UV irradiation in the presence of CO_2 and H_2O, two peaks were observed due to the C–H stretching vibration of the CH_x species and the O–H stretching of the surface hydroxyl groups at around 2920 cm^{-1} and 3630 cm^{-1}, respectively. On the other hand, only a weak peak assigned to the O–H stretching vibration was observed without UV irradiation, suggesting that UV-light irradiation is indispensable for attaining CO_2 reduction and the formation of the active H and CH_x species.

Defect of oxygen vacancy: TiO_2 is a nonstoichiometric compound, which contains defects oxygen vacancies (V_O) and titanium interstitials.[111]

These defects played essential role in photocatalytic degradation of organics and water splitting by TiO_2.[111–114] Three reasons are proposed for relating the importance of defects in TiO_2. First, the creation of defects could introduce intermediate surface states that narrow the bandgap and hence extending the photoresponse of TiO_2 to visible-light range.[115] Second, the donor state located below the TiO_2 CB promotes the charge separation and transfer.[116] Third, the formed Ti^{3+} and V_O are considered to be important active sites for the adsorption and activation of reactants.[117] As illustrated in Fig. 11 electrons could be excited from the VB to the defects donor state (red dashed line) under visible-light illumination. The lifetime of electrons in the defects donor state is much longer than that on the CB, which facilitates the formation of superoxide radicals by attachment of electrons to oxygen.[118] Meantime, the holes left in the VB accelerate the generation of free OH radicals. Both radicals are responsible for the enhanced activity of TiO_{2-x} for the photodegradation of organic compounds.[119]

Although oxygen-deficient TiO_{2-x} is a promising candidate as a photocatalyst, only limited work has been done for photocatalytic CO_2 reduction. DeSario *et al.* fabricated defective anatase/rutile thin film by direct current magnetron sputtering and investigated its CO_2 photoreduction.[115] They demonstrated that the photocatalytic activity was influenced by the levels of oxygen deficiency, which were controlled by adjusting the oxygen partial pressure during the film deposition process. They also suggested that V_O likely had two competing roles: (1) V_O could enhance

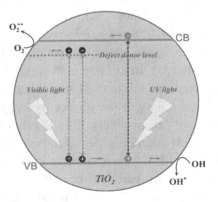

Fig. 11. Schematic illustration of charge transfer mechanism in defective TiO_{2-x}.

visible-light harvesting and act as active/adsorption sites, and (2) V_O may serve as or induce the creation of active interfacial sites at certain concentrations. Beyond that level these sites may function as recombination centers to hinder the photoactivity. Further studies are necessary to develop more stable and more efficient TiO_{2-x} for solar fuels production, and to better understand the specific roles of defects (i.e. VO, Ti interstitial) and the dynamics of charges separation and transfer.

Besides the physical aspects, the photocatalytic system can be influenced significantly by the reaction condition such as the ratio of reactants $H_2O:CO_2$, the amount of TiO_2 catalyst used, reaction temperature, light source and so on.[95] This chapter focuses on the natural characteristics of photocatalysts, therefore we don't consider much about this aspect, which can be referred the available reviews.[59,120,121]

5. Non-titanium Containing Solids for Photocatalytic CO_2 Reduction

Despite the great advances made in the photocatalytic CO_2 reduction with TiO_2 and derivative materials, the efficiency of the process still remains low. Examples of the non-TiO_2 photocatalytic systems are collected and described in detail in the following sections on sulfide, oxide, and phosphide semiconductors.

5.1 Metal oxide

Some prospective types of metal oxide and mixed metal oxide semiconductors have been reported, including ZnO, ZrO_2, Ga_2O_3, Ta_2O_5, $SrTiO_3$, $CaFe_2O_4$, $NaNbO_3$, $ZnGa_2O_4$, Zn_2GeO_4 and $BaLa_4Ti_4O_{15}$, etc.[122,123] As most of them have a large bandgap, which can provide great overpotential for the reduction, these metal oxides usually are UV active.

In addition to TiO_2, ZnO is probably the second-most studied semiconductor and has been tested as photocatalyst in a wide range of conditions.[124,125,126] As expected, the photoactivity of ZnO depended on its microstructure, crystal size, morphology and crystal orientation among other variables as well. The first approach of CO_2 conversion to carbonaceous fuel by using metal oxide catalysts was established by Inoue et al.[56]

It has been studied on three metal oxides (TiO_2, ZnO and WO_3) to understand their photocatalytic characteristics under UV light radiation. In this context, WO_3 does not respond to the photocatalytic CO_2 reduction and shows zero yield because of more negative redox potential of solution species with respect to the conduction band energy level. Another comparison of the efficiency of the TiO_2, ZnO, and NiO semiconductors for the photocatalytic reduction of CO_2 to CH_3OH in water was carried out in a study for which 355 nm laser irradiation has been used.[127] According to Fig. 12, the CH_3OH production reaches a maximum at 90 min when NiO is used; this is in contrast to the preferential use of TiO_2, which exhibits the lowest CH_3OH production at different times. In addition to CH_3OH, trace amounts of CH_3COOH in the liquid phase and CH_4 and CO in the gas phase were also detected. All these metal oxide photocatalysts with d^{10} configurations (In^{3+}, Ga^{3+}, Ge^{4+}, Sn^{4+}, Sb^{5+}) have attracted considerable attention as hybridization of the s- and p-orbitals of these metals in the conduction band leads typically to a high mobility of the photogenerated electrons, a desirable property that should enhance photocatalytic activity.[128–130] Typically, gallium oxide (Ga_2O_3) is a promising CO_2 reduction photocatalyst because of its high reduction potential (−1.449 V vs. NHE at pH 7) that meets the required thermodynamic value for CO_2 reduction, that is, $E_0(CO_2/CH_4)$= −0.24 V and $E_0(CO_2/CO)$ = −0.53 V(vs. NHE at pH 7).[131–136]

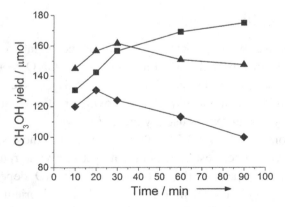

Fig. 12. CH_3OH yield during the photoreduction of CO_2 using 355 nm laser irradiation over TiO_2 (◆), NiO (■), and ZnO (▲) as photocatalysts.

The pre-synthesized $InTaO_4$ by solid-state synthesis starting from In_2O_3 and Ta_2O_5 was impregnated with 0–1.0 wt.% NiO using an aqueous nickel nitrate solution, calcined at 623 K, reduced at 773 K, oxidized at 473 K, and then used as a catalyst for the photcatalytic CO_2 reduction.[137] It was found that methanol was formed at a maximum rate of 1.4 μmol $h^{-1}g^{-1}$ in a CO_2-saturated potassium bicarbonate solution using $NiO/InTaO_4$ catalyst. Unfortunately, the contrast reactions (in the dark, in the absence of catalyst, in the absence of CO_2 and/or $KHCO_3$) and additional product information, including analysis of the gas phase were not reported. The $NiO/InTaO_4$ photocatalysts were also tested for the gas-phase CO_2 reduction using a monolith reactor.[138] The major product was acetaldehyde, which was formed at a rate of 0.21 μmol $h^{-1}g^{-1}$, rather than minor methanol formation.

$InNbO_4$ was also synthesized by solid-state reactions and then tested for the CO_2 photoreduction in a $KHCO_3$ solution.[139] Based on the control tests and the fact that the catalyst was prepared via a carbon-free synthetic method, the methanol formation, which occurred at a rate of 1.3 μmol $h^{-1}g^{-1}$ under UV–visible-light, should be photocatalytically active. Hydrothermally synthesized HNb_3O_8 nanobelts also underwent the photoreduction of CO_2 to methane with rates of 3.6 μmol $h^{-1}g^{-1}$, which are comparable to TiO_2-based photocatalysts. Pt-promoted perovskite $NaNbO_3$ was reported to produce methane from $CO_2 + H_2O$ at a rate comparable to the levels of the other best semiconductor photocatalysts: 4.9 μmol $h^{-1}g^{-1}$.[140-142]

$ALa_4Ti_4O_{15}$ (A=Ca, Sr, or Ba) with a layered perovskite structure and bandgap of 3.79–3.85 eV were synthesized by using different methods and used for CO_2 photoreduction in aqueous solutions.[143] Iizuka et al. found that $ALa_4Ti_4O_{15}$ can reduce water without additional reductant. However, hydrogen was the preferential reduction product. Out of the various photocatalysts tested, $BaLa_4Ti_4O_{15}$ was determined to be the most active photocatalyst and its selectivity was shifted toward CO_2 reduction after introducing Ag metals on the surface. The optimal loading amount was determined to be 2.0 wt.% Ag-loaded $BaLa_4Ti_4O_{15}$, with maximum H_2, O_2, CO and HCOOH yields of 10 μmol h^{-1}, 16 μmol h^{-1}, 22 μmol h^{-1} and 0.7 μmol h^{-1} respectively.

AB_2O_4-type oxides with spinel structure have received considerable attention in catalysis, as gas sensors, and in photoelectronics. Specifically,

$ZnGa_2O_4$ is a transparent and conductive material that can be useful for UV photoelectronic devices and finds applications in wastewater treatment.[144–146] Mesoporous $ZnGa_2O_4$ was prepared through ion exchange starting from mesoporous $NaGaO_2$ as precursor, and its photocatalytic activity for CO_2 photoreduction was tested.[147,148]

Recently, many other oxides with visiable light response have been fabricated through various routes and used in the photochemical CO_2 reduction. For example, microwave assisted hydrothermally-derived monoclinic $BiVO_4$ shows higher yield (110 $\mu molh^{-1}g^{-1}$) of ethanol compared to tetragonal $BiVO_4$, due to CO_3^{2-} being anchored to the Bi^{3+} sites on the external surface through a weak Bi–O bond to receive the photogenerated electrons effectively from the V 3d-block bands of $BiVO_4$. The asymmetric behavior of monoclinic phase around the Bi^{3+} ion is more effective than in the tetragonal phase which is the main reason why Bi^{3+} ion shows stronger lone pair behavior in the monoclinic phase to increase the tendency of Bi–O bond formation with CO_3^{2-}.[125,149]

In various types of photocatalysts, the metal oxides are more preferable, due to their potential advantages, e.g. being relatively safe to handle, fairly low cost and considerable stability, thus particular attention is still paid to photoreduction of CO_2 using metal oxide photocatalysts.[150,151] Actually, there are many other different types of metal oxide photocatalysts which are commonly studied in various investigations, and here we cannot list them one by one.

5.2 Metal sulfide

From a historical perspective, metal sulfide semiconductors, apart from the metal oxide semiconductor catalysts, have effective photoactivity because of their outstanding ability to absorb the solar spectrum with high energy yields and received a lot of attention.[152] Metal sulfides have comparatively high conduction band states and are more appropriate for enhanced solar responses than the oxide analogues, due to the higher valence band states consisting of S 3p orbitals. Many sulfides as photocatalysts for CO_2 reduction have a narrow bandgap (e.g. PbS, Bi_2S_3, CdS, Bi_2S_3/CdS, $Cu_xAg_yIn_zZn_kS_m$ etc.), with the onset of absorption in the visible or even infrared region.

ZnS and CdS were the most studied metal sulfides for photocatalytic CO_2 reduction. ZnS is a typical direct wide bandgap semiconductor (Eg = 3.66 eV in the bulk), which absorbs only in the UV range. However, it possesses a strongly reducing conduction band (ECB = −1.85 V vs. the NHE at pH 7),[153] and has been reported to preferentially reduce CO_2 to formic acid with high quantum yields.[154–157] Henglein et al.[155] and Yoneyama[158] speculated that smaller ZnS NPs are more efficient due to their larger bandgap, higher surface area, and greater affinity for CO_2. Fujiwara investigated ZnS NPs with a diameter of 2 nm in DMF for the photocatalytic CO_2 reduction when zinc acetate was used as a precursor of the nanocrystals.[154] It was found that the close interaction of acetate ions to Zn atoms inhibited the creation of sulfur vacancies as catalytic sites for CO production. In addition, surface dimethylformamide (DMF) solvated Zn atoms provide more efficiency to $HCOO^-$ formation. On the other hand, the addition of excess Zn^{2+} enhances the rate of formation of HCOOH through the reaction between excess Zn^{2+} and the surface of ZnS–DMF(OAc) nanocrystallites.

As a consequence of the narrow bandgap of 2.4 eV, the onset of absorption with CdS is at 520 nm.[159] However, the conduction band electrons of CdS are less reductive (ECB = −0.9 V at pH 7 vs. the NHE) compared to those of ZnS.[160] CdS nanoparticles with and without surface modification by thiols have been used as photocatalyst for CO_2 photoreduction in various solvents.[161] The main photoproducts observed during the CO_2 photoreduction using thiol-capped commercial CdS were formate, CO and H_2. Fig. 13 shows the fraction of CO_2 reduction products obtained by irradiation of various surface-modified CdS photocatalysts as a function of the solvent containing 2-propanol as sacrificial electron donor. The significance of these results is to illustrate how the nature of the semiconductor surface can influence the product distribution. The use of solvents with a different dielectric constant of DMF, CH_3CN, H_2O etc. varies the ratio of formic acid to CO.[154,161]

Other sulfides were rarely used as photocatalysts. Manganese sulfide (Eg = 3.0 eV) reduced bicarbonate dissolved in water (pH 7.5) to formic acid with a photonic efficiency of 4.2% under UV irradiation.[162] With a narrow bandgap of 1.28 eV, Bi_2S_3 absorbs in the visible range and has been shown to reduce CO_2 to methanol, especially when used in

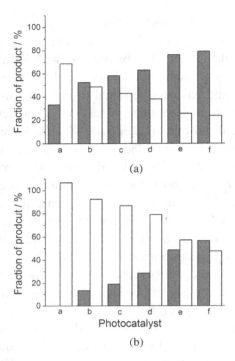

Fig. 13. Percentage of HCOOH and CO (&) from CO_2 photoreduction using various CdS photocatalysts for 7 h in acetonitrile (A) or dichloromethane (B). CdS (a); CdS surface modified with 2-aminoethanethiol (b), 1-dodecanethiol (c), 2-mercaptoethane sulfonate (d), sodium sulfide (e), and 0.5m 1-docecanethiol (f). The total sum of the photoproducts was less than 2 mmol in 7 h.

conjunction with CdS particles.[150] 15 wt.% Bi_2S_3 performs the highest formation rate (88 μmolg^{-1}h^{-1}), which was two to three times higher than for either semiconductor alone. A limitation of using metal sulfides in photocatalysis is that they are generally not stable under illumination in an aqueous dispersion because of the oxidation of lattice S^{2-} ions to elemental sulfur and eventually to sulfates. To prevent the photocorrosion, additional reducing agents such as sulfite (SO_3^{2-}), thiosulfate ($S_2O_3^{2-}$), or hypophosphite ($H_2PO_2^-$) anions, tertiary amines (e.g. triethylamine, TEA), or alcohols (e.g. propan-2-ol) were used to scavenge the photogenerated holes.

5.3 Phosphide

Several main group metal phosphides, especially GaP and InP with narrow bandgaps, have also been investigated for their ability to reduce CO_2 in photochemical or photoelectrochemical setups.[36,163] GaP is a p-type semiconductor with a reasonably narrow bandgap (2.3 eV) and a very high reducing position of the conduction band. It is used mainly in methanol production but need remarkably high over-potentials.[164] Barton et al. found that p-type GaP was able to obtain faradaic efficiency of 100% while converting CO_2 to methanol at low potentials, which was more than 300 mV below the standard potential of −0.5 V vs. the saturated calomel electrode (SCE). Habisreutinger et al. reported that p-type GaP semiconductor exhibits low bandgap energy (2.24 eV) and possesses high reducing conduction band electrons to ease the reduction of CO_2.[165] Pyridine was developed to act as a co-catalyst and was thought to be responsible for the selectivity of the process towards methanol through intermediate carbamate species.

P-type InP semiconductors tend to photoreduce CO_2 to formic acid selectively, although the minimum of its conduction band is located about 0.85 V lower than that of GaP.[166] Moreover, Kaneco et al. studied the photoelectrochemical reduction of CO_2 based on p-type InP, which revealed higher positive potential of 0.2–0.4 V than that of p-type Si and p-type GaAs photocathodes in the electrolysis of methanol electrolyte.[163] Arai et al reported a zinc-doped InP photocathode (Eg = 1.35 eV) modified with a ruthenium coordination polymer $[Ru(L–L)(CO)_2]_n$, in which L–L refers to a diimine ligand.[167] In an aqueous environment, the system exhibited good selectivity for the formation of formate under irradiation with visible-light (λ>400 nm) and an applied potential of −0.6 V. Importantly, it has been proven by the isotope analysis using $^{13}CO_2$ and D_2O that the CO_2 and water act as sources of carbon and protons, respectively. The system was further combined with a second photocatalyst based on TiO_2–Pt. This design was utilized to oxidize water in a separate compartment.[168] The so called Z-scheme system is composed of two photocatalysts, which have similar bandgap widths but shifted band edges. The shifting happens based on a two-step excitation, where the less reductive electron recombines with the less oxidative hole, leaving the more active species to drive redox reactions. In such a scheme, InP modified with a ruthenium electrocatalyst photoreduced CO_2 to formate without any external bias, with a power conversion efficiency of 0.03%, thus underlining the viability of the two-compartment photoelectrochemical approach.

5.4 Solid solutions

The solid solutions exhibit better performance than the single components comprising the solid solution. Solid solutions such as $(Ga_{1-x}Zn_x)(N_{1-x}O_x)$,[169] $(Zn_{1+x}Ge)(N_2O_x)$[170,171] and zinc gallogermanate solid solution[172] was utilized for photocatalytic reduction of CO_2. Sheaf-like, hyperbranched Zn_2GeO_4 nanoarchitectures have been prepared using a solvothermal route in a binary ethylenediamine/water solvent followed by nitridation under NH_3 flow to yield a yellow zinc germanium oxynitride ($Zn_{1.7}GeN_{1.8}O$) solid solution.[170] The material modified with 1 wt.% Pt and 1 wt.% RuO_2 shows a higher activity for CO_2 reduction with apparent quantum yield of 0.024% at wavelength of 420 ±15 nm (Fig. 14). The p–d repulsion of N2p and Zn3d orbitals lifts the top of VB and then narrows the bandgap. Simultaneously, the introduced Pt and RuO_2 can trap the electrons and holes, respectively, hence inhibiting the recombination of electrons and holes. Incorporating Zn_2GeO_4 into $ZnGa_2O_4$ to form Zinc

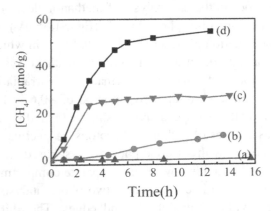

Fig. 14. CH_4 generation over $Zn_{1.7}GeN_{1.8}O$ (▲), 1 wt% Pt-loaded $Zn_{1.7}GeN_{1.8}O$ (o), 1 wt% RuO_2-loaded $Zn_{1.7}GeN_{1.8}O$ (♦), and 1 wt% RuO_2+1 wt% Pt co-loaded $Zn_{1.7}GeN_{1.8}O$ (■) as a function of visible-light irradiation time.

gallogermanate (denoted as 4.5($ZnGa_2O_4$):(Zn_2GeO_4)) can narrow bandgap by the upshift of VB edge resulting from the increased p–d ($O2p$-$Zn3d$) repulsion effect through incorporation of s and p orbitals of Ge, and the downshift of CB edge by introducing the low-energy orbital of Ge.[172] The zinc gallogermanate solid solution possesses a light-hole effective mass which is beneficial for improving hole mobility, and thus enhances the ability for water oxidation to provide protons for CO_2 reduction.

A ZnS–$AgInS_2$–$CuInS_2$ solid solution with a bandgap < 2 eV, which is dependent on the compositions, was capable to have an absorption edge of up to 800 nm.[173] The highest formation rate of methanol in the presence of sodium nitrite with 34.3 $\mu molg^{-1}h^{-1}$ was observed over $Cu_{0.12}Ag_{0.30}In_{0.38}Zn_{1.22}S_2$. Recently Arai et al. employed Cu_2ZnSnS_4, a quaternary solid solution with a narrow bandgap (Eg = 1.5 eV) and reductive conduction band (ECB = 1.3 V), to photoelectrochemically reduce CO_2 to formate in the presence of a metal complex electrocatalyst.[174] This semiconductor operated effectively under illumination with visible-light, and its p-type character facilitated electron transfer to the catalyst, where CO_2 reduction proceeded preferentially to proton reduction. Nevertheless, it required an external bias of 0.4 V, in place of a hole scavenger, to prevent photooxidation.

5.5 MOFs

As a new family of inorganic–organic hybrid supramolecular materials, MOFs serve as an interesting platform to design chemical functions, such as absorption and catalysis. As the ligand-to-metal charge transfer (LMCT) and ligand p–p^* excitation of MOFs fall into UV and the blue end of visible regions, MOFs have been proposed to exhibit similar photocatalytic activities as semiconductors. Meanwhile, MOFs can in principle contain photosensitizers and catalytic centers in a single solid and provide the structural organization to integrate the three fundamental steps of artificial photosynthesis into a single material. Different molecular complexes were doped into Uio-67(Zr) to create the first MOF-based heterogeneous catalytic system. The novel system was developed for water oxidation, photocatalytic CO_2 reduction, and organic transformations under visible-light irradiation, which proves the potential of utilizing functionalized MOFs as a photocatalysts for CO_2 reduction under visible-light. (Fig. 15).[175,176]

Although photoreduction reactions can be achieved with these simple MOFs, the activities are very low, which is probably due to the absence of vacant sites for substrate activation and inefficient electron transfer between the MOF and the substrate. To enhance their performances, several MOFs have been designed and used as either photosensitizers or catalysts in multi-component photocatalytic systems. Lin and coworkers reported the synthesis of a MOF photocatalyst $Zr_6(\mu_3\text{-O})_4(\mu_3\text{-OH})_4(\text{bpdc})_{5.83}(L_8)_{0.17}$ (bpdc = 5,5'-biphenyldicarboxylate; H_2L_8 = $Re(CO)_3(5,5'\text{-dcbpy})Cl$) by doping $[Re(CO)_3(5,5'\text{-dcbpy})Cl]$ into a Uio-67 framework. $Re(bpy)(CO)_3X$ has been extensively studied as a molecular CO_2 reduction catalyst in homogeneous systems. It showed the same powder X-ray diffraction pattern as UiO-67 due to the matching lengths of $[Re(CO)_3(5,5'\text{-dcbpy})Cl]$ and 4,40-biphenyldicarboxylic acid. Highly selective photocatalytic CO_2 reduction towards CO in acetonitrile solution with triethylamine as a sacrificial reducing agent was achieved using this MOF catalyst.

Wang and coworkers reported the photocatalytic activity of a cobalt-based zeolitic imidazolate framework $Co(\text{benzimidazole})_2$ (Co–ZIF-9). Co–ZIF-9 showed photocatalytic activities towards proton

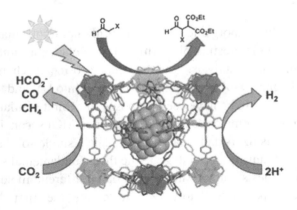

Fig. 15. Schematic illustration of MOF-based heterogeneous catalytic system for water oxidation, photocatalytic CO_2 reduction, and organic transformations.

and CO_2 reduction in the presence of the $Ru(bpy)_3^{2+}$ photosensitizer, the TEOA sacrificial reducing agent and visible-light. A mixed solution of Co^{2+} and the benzimidazole ligand also showed some catalytic activity, but lower than that of Co–ZIF-9. However, the selectivity of CO_2 reduction against proton reduction is moderate. Controlled experiments indicated that the efficiency for proton reduction increased greatly in the presence of CO_2, but it remains unclear whether it is caused by pH difference or CO_2 playing a role in the hydrogen evolution catalytic cycle.

For most MOFs, a second component, either a photosensitizer or a co-catalyst, is needed to get enhanced efficiency for proton or CO_2 reduction. In these cases, slow mass transport in MOF channels and inefficient electron transfer between MOF and the homogeneous component will greatly decrease the activity. Rational designs are thus needed to integrate different components in a single solid to obtain a more active catalyst.

5.6 Carbon-based photocatalyst

More recently, nanocarbons, such as fullerene,[177] nanodiamond,[178] carbon nanotube (CNT),[179] and graphene oxide (GO)/graphene,[180,181] exhibiting high thermal conductivity, high theoretical specific surface area, unique carrier mobility, low-dimensional structure, and sp^2-hybridized carbon configuration, have demonstrated their promising functions as promoters or catalysts for photorecations. Extensive studies have used CNT, graphene, and GO as supports or promoters for enhanced production of solar fuels.

Using CNT hybrid photocatalysts as an example, a large number of studies have been addressed for photodegradation of organic pollutants,[181-183] yet very few reports related to CO_2 reduction. In order to be applied in the reduction of CO_2 with H_2O, MWCNT-suported anatase TiO_2 nanoparticles and rutile TiO_2 nanorods were prepared by sol-gel and hydrothermal methods, respectively.[184] A proper amount of MWCNTs is used in the system in order to reduce the agglomeration of TiO_2 and it can also effectively transport the photoinduced electron–hole pairs along the tubes, by which the recombination rate of carriers is decreased and the photocatalytic activity is enhanced. However, excessive amount of MWCNTs can lead to less photoinduced carries for the CO_2 reduction by shielding TiO_2. The differences in the main products of C_2H_5OH over anatase and HCOOH over rutile were observed and attributed to the varying crystal phases of titania. In addition, the extension of the visible-light absorption can be achieved by CNT modification which can lower the bandgap energy and improves the electronic structure of TiO_2.[185] The excited electrons from TiO_2 can transport to the conductive structure of CNTs, because they have a lower Fermi level via formation of a Schottky barrier at the interface. The heterojunction of CNTs and TiO_2 is also beneficial to charge separation, stabilization, and hindered recombination.

As a honeycomb structured carbon sheet with sp^2 bonding, graphene has unique electronic properties, large theoretical specific surface area, and high transparency. These figures of merit make graphene an excellent candidate for modifying a photocatalyst for enhanced photocatalytic performance.[180,186,187] Graphene hybridized photocatalysts have been studied intensively for CO_2 reduction in the past several years. In TiO_2–graphene 2D sandwich-like hybrid nanosheets abundant Ti^{3+} was detected on the surface of TiO_2.[188] CH_4 and C_2H_6 were produced by photoreduction of CO_2 over the hybrid (Fig. 16). The total production rate was boosted with the increasing content of graphene. The d orbital of TiO_2 and π orbital of graphene can match well in energy levels and form a d–π electron orbital overlap. Therefore, the photoinduced electrons from TiO_2 can transfer freely along the graphene network for CO_2 reduction, increasing the photocatalytic CO_2 conversion rate.

Other metal-free carbonaceous materials of modified graphene, graphene–CNT,[189] graphene–C_3N_4,[190] and CNT–C_3N_4[191] have demonstrated their usages in photocatalysis as well. More recently, photocatalytic CO_2 conversion to methanol was investigated over GO as a metal-free photocatalyst. The results show the photocatalytic methanol formation over GO-1 (prepared by Hummer's method), GO-2 (modified with 5 mL of H_3PO_4), GO-3 (modified with 10 mL of H_3PO_4), and P25. Activity of methanol formation from photocatalytic CO_2 conversion followed the order: GO-3 > GO-1 > GO-2 > P25. The photocatalytic

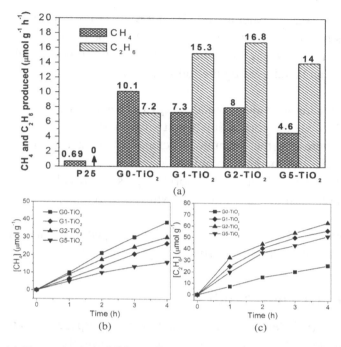

Fig. 16. (a) Photoreduction of CO_2 on G_x–TiO_2 (x = 0, 1, 2, and 5) and P25. (b and c) Photocatalytic CH_4 and C_2H_6 evolution amounts for samples G_x–TiO_2 (x = 0, 1, 2, and 5).

CO_2 to methanol conversion rate of the modified GO-3 was six times higher than that of P25.

6. Strategies for Enhanced Photocatalytic CO_2 Reduction

As has been demonstrated, the current state-of-the-art has shown the feasibility of photocatalytic CO_2 reduction, although it is far from reaching efficiencies for practical application. The rate of the formation of products is usually in the μmol. g^{-1}. h^{-1} scale. Some of the key factors that limit the efficiency are listed: (1) mismatch between the absorption ability of semiconductor and the solar spectrum; (2) poor charge carrier separation efficiency; (3) low solubility of CO_2 molecule in water (approximately

33 μmol in 1 mL of water at 100 kPa and room temperature); (4) reversed reactions during reduction of CO_2; and (5) competition reaction of water reduction to hydrogen. Specifically, the quest for new semiconductor materials can be advised by the following points:[192]

1) Raising the valence band energy to decrease the bandgap;
2) Moving the conduction band to more reductive potentials;
3) Improving the quantum efficiency of excitation formation and reducing charge recombination;
4) Using novel nanoscale morphologies to provide a large surface area with multiple photocatalytically active sites.

Several reliable strategies have been developed to enhance the photocatalytic CO_2 reduction properties.

6.1 Doped photocatalysts

Doping external elements is deliberately for the purpose of controlling its electrical properties. The VB level of most currently developed metal oxide photocatalysts is normally located about 3 eV vs. NHE,[193] which confines the metal oxide only to the UV-reduction of CO_2 reduction. To engineer its bandgap, surface modification of the photocatalysts has been attempted by introducing oxygen vacancies in semiconductors,[194] or incorporating foreign elements.[195–198] The schematically photocatalytic view of a doped semiconductor is illustrated in Fig. 17.

Non-metal doped: Doping or co-doping of semiconductor photocatalysts with non-metals such as C, N, S, B and F has considerable effect on engineering low bandgap energy, leading to efficient photocatalytic activity under visible-light irradiation. N-doped mesoporous TiO_2 with Pt, Au, and Ag as co-catalysts absorbed light up to 500 nm in the visible-light spectrum, which contributes to photocatalytic visible-light CO_2 reduction ($\lambda > 420$ nm).[199] N-doped Ta_2O_5 linked with Ru complex electrocatalysts showed selective conversion of CO_2 to HCOOH under visible-light irradiation (410 nm $< \lambda <$ 750 nm).[200,201] N-doped $InTaO_4$ with Ni@NiO core shell nanostructure as co-catalyst displayed photocatalytic reduction of CO_2 with H_2O into CH_3OH under light irradiation (390 nm $< \lambda <$ 770 nm).[202] The

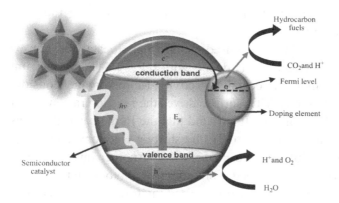

Fig. 17. Photocatalytic reaction mechanism of doped semiconductor.

investigation with carbon-doped TiO_2 was done by Xue et al. In their work, the doped carbon narrows the bandgap, shifts the absorption to the visible-light region and facilitates the charge separation efficiency. Hiroshi et al. reported that the hydrophilic behavior of C-doped TiO_2 semiconductor exhibited visible-light response, which was initiated from the localized C (2p) formed above the valence band.[203] S-doped TiO_2 has the ability to reduce the bandgap energy and inhibits the alteration of anatase phase into rutile phase at elevated temperatures.[204] Correspondingly, the doping of F on a semiconductor catalyst performs better by decreasing the bandgap energy.[205] More considerably, iodine doped TiO_2 catalysts exhibited higher photochemical activity due to the closer ionic radii of I^{5+} and Ti^{4+} for substituting lattice titanium.[206,207] The p–n hybrid system of hollow CuO nanocubes decorated with nanoparticles of titanium oxynitride ($TiO_{2-x}N_x$) was reported to photocatalytically convert CO_2 into CH_4 at ambient temperatures.[208]

Metal doped: Doping metals to semiconductors can act as an electron trap to provide electrons to reduce CO_2. The photoexcited electrons are shifted from the conduction band of the semiconductor to metals to leave the holes on the surface of semiconductor.[209] Noble metals (including Ru, Rh, Pd, Ag, Pt, Cu, Ni and Au),[143,210–221] transition metals (including Co, Mn, Ni, Cr, Mo, Fe, V, and Ti)[222–225] and co-metals (including Fe–Cu, Cu–Pt, Ag–Pt et al.)[226–230] have been commonly used with semiconductors as dopants due to Fermi level or electron-accepting areas at the energy just beneath the conduction band of semiconductors for the photochemical reduction of CO_2.

Pt is one of the most used metals in photocatalysis. For example, the TiO_2 semiconductor was doped with 0.5 wt.% Pt by wet impregnation presenting high photocatalytic activity. More homogeneous dispersion of 0.5 wt.% Pt on the TiO_2 surface drives the reaction rate swiftly in the both liquid and gas phases. It is found that the Pt mediated the transformation of active sites from Ti^{4+} to Ti^{3+} by shifting electrons, and this phenomena facilitates visible-light response, too.[231] Another investigation shows that Pd doped TiO_2 performed more effective photocatalytic activity than that of other metals, and it also increased the formation of CH_4 instead of CO from CO_2.[212] On the other hand, Pd on the surface of TiO_2 partially oxidized to form PdO during the progression of CH_4 formation, which led to the deactivation of the Pd–TiO_2 catalysts.[62] Doping with proper metals is also a simple and effective practice to shift the band edge from the UV region to the visible region based on the plasmon effect.[219–221]

The experiments have been done to characterize the different types of transition metals (V, Fe, Ce, Cu, Cr) doped TiO_2.[225] It was reported that V and Fe were located in the substitutional sites of TiO_2, and Ce ions were dispersed in the interstitial sites, although Cr and Cu were accumulated on the surface. The activity order was found in the following sequence: Fe–TiO_2 > V–TiO_2 > Cr–TiO_2 > Ce–TiO_2 > TiO_2 > Cu–TiO_2, which indicates that the local structure and type of dopant have vital contributions on the photocatalytic technology. As usual, a suitable percentage of metal doping could improve catalytic activity, whereas excess percentages of dopant metal decreased the photocatalytic performances because of the occurrence of charge recombination.

Suitable doping of bimetallic elements together onto semiconductors has been validated as a potential method to advance the visible-light response of photocatalysts. Bimetallic Ag–Pt were incorporated into the TiO_2 as reported by Mankidy et al. They obtained a higher selectivity of 80% for CH_4 than 20% with single TiO_2.[230] Their investigation showed that Ag species had a strong surface plasmon absorption band in the UV-vis region, while the Pt species did not show that behavior. However, the combination of bimetallic Ag–Pt has the ability to tune the electronic properties during photochemical reactions.

6.2 Heterostructures

Heterostructures consisting of two or more materials can optimize charge separation and photocatalytic properties by providing an additional control of the electron and hole wave functions (i.e. wave function engineering).[232] Both semiconductors in the heterojunctions are photoexcited to generate e^-h^+ pairs. The energy bias between the two sides facilitates the electrons and holes transfer between them. The electrons could transport from the semiconductor with a higher CB to the one with a lower CB. Meanwhile, the holes could transport from the semiconductor with a lower VB to the one with a higher VB. In this aspect, the separation and transfer of charge carriers are enhanced, leading to higher photocatalytic efficiency.

Xi *et al.* (2011) developed a nanoporous "French fires" shaped TiO_2–ZnO composite, which demonstrated a six times higher activity than P25 for CO_2 photoreduction. Figure 18 shows the charge transfer process in the TiO_2–ZnO heterojunction for CO_2 photoreduction. The photogenerated electrons migrated from ZnO CB to TiO_2 CB, since the CB edge of ZnO (–0.31 V) was higher than that of TiO_2 (–0.29 V). Meanwhile, the holes migrated from TiO_2 VB (2.91 V vs. NHE) to ZnO VB (2.89 V vs. NHE).

Other semiconductors such as CuO, Cu_2O, and $FeTiO_3$ have been paired with TiO_2 for CO_2 photoreduction as well.[1,208,233–235] These semiconductors have a narrow bandgap (Eg = 1.8–2.5 eV) and relatively high

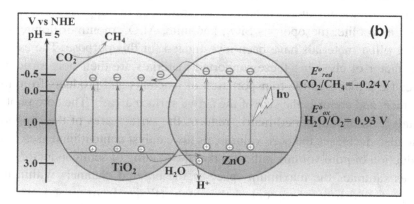

Fig. 18. Schematic illustration of the band structure and charge separation in the hybrid TiO_2/ZnO composites.

absorption coefficient in the visible region. Consequently, the heterojunctions formed between these semiconductors and TiO_2 may enhance the separation of charge carriers and the response of visible-light simultaneously. The incorporated CuO and Cu_2O, can act as electron trapping sites to enhance electron transfer from TiO_2 to CuO_x. A key factor that determines the function of CuO_x in a CuO_x–TiO_2 composite catalyst system is the concentration and crystal structure of CuO_x. If the concentration of CuO_x is small (e.g. 0.1 to 2%), it exists as amorphous and highly dispersed on the surface of TiO_2. In this case, the dispersed Cu^{2+} or Cu^+ species trap electrons from TiO_2 undergoing redox processes of Cu^{2+}/Cu^+ or Cu^+/Cu^0.[236] When the concentration of CuO_x is very high (close or more than the TiO_2), the crystalline particles forms and behave as semiconductor. In this case, the heterojunction effect plays a more primary role. As the CB edge of TiO_2 is close to the VB edge of CuO, photoexcited electrons in the TiO_2 CB could recombine with holes produced in the CuO VB, while higher energy electrons left in the CuO CB reduce CO_2 to hydrocarbons and the holes left in the TiO_2 VB oxidize H_2O to O_2. This charge transfer mechanism at the CuO_x/TiO_2 heterojunction is different from that at the ZnO/TiO_2 heterojunction as previous described.

6.3 Supported materials for catalyst immobilization

The photocatalytic conversion efficiency is also dependent on different supports/immobilization materials. Several types of supported materials, such as zeolite, mesoporous SiO_2, kaolinite, Al_2O_3, montmorillonite and other pillar materials have been investigated for this purpose. The essential factor of choosing these supported materials are their pore structure because it manages the transportation of reactants and products from the surface, and controls the size of the active surface area.[237] The core photocatalyst on the supported matrix reduces the surface area of the supports and also decreases the pore volume, as the catalyst content increases. The reduction of pore volume indicates that the loading of semiconductor particles saturates the maximum space of the ordered channels within the support, which helps to control the catalysts' size from crystal growth. High catalyst loading can cause pore blockage inside the channels, although the interconnection among the parallel pore channels should

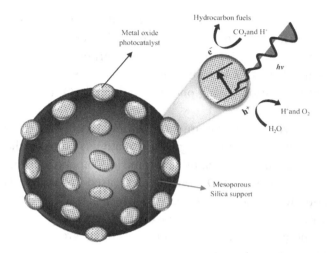

Fig. 19. Photocatalytic reaction mechanism of supported metal oxide semiconductor.

have the ability to reduce this effect.[238] The common photocatalytic reaction path for supported metal oxide semiconductor is illustrated in Fig. 19.

Transparent porous silica thin films are extraordinary supporting materials for TiO_2 photocatalyst since the thin films have a larger surface area in contrast to metal oxide thin films on a quartz substrate and can realize the efficient absorption of light.[239–241] The Ti-containing porous silica (Ti–PS) thin films in the presence of CO_2, H_2O and UV irradiation led to the formation of CH_4, CH_3OH as well as CO and O_2 as minor products. From FTIR analysis, the concentration of the surface •OH groups was found to play a role in the selectivity towards the formation of CH_3OH. The methanol yield over the transparent Ti–PS thin films with hexagonal pore structure is 7.5 μmolg^{-1}h^{-1} and 2.0 μmolg^{-1}h^{-1} for CH_4.[239,242]

Zeolites are microporous, aluminosilicate minerals having porous structures. The investigations show that the TiO_2 particles are highly distributed inside the pore channels of zeolite with tetrahedral coordination, which has high selectivity for the formation of CH_3OH.[243] At the same time, it is also found that Pt/TiO_2 catalysts in the presence of zeolite support facilitates the charge separation and increases the formation of CH_4 in lieu of CH_3OH by stimulating the reaction between the carbon radicals and H atoms formed on the Pt metals to produce CH_4.

Recently, montmorillonite has been loaded with TiO_2 structures to increase the surface area of TiO_2 particles.[244] The reducing of the particle size from 18.73 nm to 13.87 nm helps create efficient charge separation. The bandgap of TiO_2 was also reduced to 3.07 eV and could permit visible-light irradiation during photoreduction of CO_2 in the presence of water vapor.

Kaolinite, which is a clay mineral with a layered silicate having one tetrahedral sheet connected through oxygen atoms to one octahedral sheet of alumina octahedral,[243] is another promising supporting material. The crystallite size of anatase TiO_2 could be decreased by kaolinite/TiO_2 composite. Kaolinite increases the effective surface area for TiO_2 by avoiding the aggregation of the TiO_2 in suspension and prevents the recombination of electron–hole pairs. It also hinders crystallite growth and decreases particle sizes from 26 nm to 18 nm.[245]

In photochemical reactions, the involvement of flexible substrates such as polyethyleneterphthalate (PET), polyethylenenaphtalate (PEN), polyethylene (PE) and polypropylene (PP) have grown in interest because they deliver a high surface area for catalysts.[246] However, the degradation of polymer substrates on solar irradiation depends on the operating period of atmospheric exposure.[247] This photodegradation is carried out via a combination of photolysis and photocatalytic oxidation.[246] The formation of volatile products due to the photodegradation is another limitation of polymeric substrates.[248]

6.4 Photosensitization

Photosensitization involves a modification of catalysts with narrow bandgap semiconductors or dye molecules to realize visible-light response through the transfer of photogenerated electrons from narrow bandgap photosensitizers to the CB of wide bandgap semiconductors (Fig. 20). Narrow bandgap photosensitizers such as CdS, Bi_2S_3,[249] CdSe,[250] PbS,[251] AgBr[252], light harvesting dyes and macrocyclic ligands have been coupled with TiO_2 for photocatalytic reduction of CO_2 under visible-light irradiation. According to Tahir and Amin, photosensitization exclusively depends on the bandgaps of sensitizers in terms of visible-light response and the high conduction band energy of sensitizers compared with the catalysts'

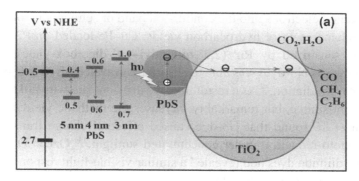

Fig. 20. Band alignment between TiO_2 and PbS QDs with different particles size, and the schematic diagram of the photocatalytic process for CO_2 photoreduction under visible-light irradiation.

conduction band energy due to facilitating electrons shifting to the catalysts.[253]

Metal chalcogenide quantum dots (QDs) have been used as photosensitizers to facilitate the photoreduction of CO_2 into hydrocarbons. Wang et al. developed a PbS QDs sensitized Cu–TiO_2 catalyst with three different sizes of PbS QDs samples (3 nm, 4 nm and 5 nm).[251] The results indicated that the maximum CO_2 transformation rate is displayed with 4 nm PbS QD-sensitized Cu–TiO_2, due to absorption of the visible spectrum ($\lambda > 610$ nm) and charge separation properties. In addition, small-size QDs inject electrons swiftly to the conduction band edge of TiO_2 and large-size QDs cover the visible region.[250] Moreover, it was found out that the electron transfer rate is very fast (less than 1 ns) from the excited PbS QDs state to TiO_2 CB state.[254] Other investigations were also conducted on the CdS, CdSe, Bi_2S_3 sensitized TiO_2 which revealed response to the visible-light region ($\lambda > 420$ nm).

Silver bromide (AgBr) was also used as a sensitizing agent due to its high sensitivity to light.[252] The system consisting of 23.2 wt.% AgBr coupled with P25 was found to have the highest yields at pH of 8.5. The detected reduction products were methane, methanol, CO and ethanol. The system exhibited almost no deterioration in performance over repeated cycles.

Organic dyes are used as sensitizers to enhance the visible-light absorption of a semiconductor too. Under irradiation, dyes can inject photoexcited electrons into the conduction band of semiconductor leading to the enhancement of hydrocarbon yields. Cu–Fe loaded on P25 photocatalysts sensitized by Ru^{II} (2,2'-bipyridyl-4,4'-dicarbox-ylate)$_2$-(NCS)$_2$ (also known as N3-dye) was able to absorb the wavelength of 400–800 nm from solar irradiation.[38] As a result, the dye does not show its full response to the light source, but remarkably increases the methane yield to 100%. Nguyen et al. found that N3-dye caused effectual charge transfer with higher photo-reduction under concentrated sunlight.[255] Ozcan et al. used Perylene diimide dyes and revealed a similar visible-light response to Ru^{II} (2,2'-bipyridyl)chloride hexahydrate.[256]

Other macrocyclic ligands, such as phthalocyanins (Pcs) and porphyrins (Prs), can transfer photogenerated electrons directly into the CB of TiO_2 as well.[219,257] Such macrocyclic ligands are good candidates for sensitization because of their high absorption coefficient in the visible region and good chemical stability. The incorporation of 0.7% CoPc into TiO_2 exhibited a high yield rate of formic acid (about 150 $\mu mol g^{-1} h^{-1}$).[258,259] The investigation conducted on ZnPc–TiO_2 found that ZnPc has a higher selectivity of formic acid than metal catalysts by decreasing the recombination possibility of hole–electron pairs.[260] It is worth noting that the excessive amount of ZnPc can reduce the photocatalytic activity, because the active TiO_2 surface sites can be blocked by additional metal phthalocyanines, and the optimum amount is 1.0 wt.% ZnPc. According to the observations of Wang et al., the bandgap of TiO_2 can be minimized from 2.95 eV (at 0.6 wt.% ZnPc) to 2.79 eV by loading 5 wt.% ZnPc.[261]

6.5 Sacrificial agent

An increase in charge carrier recombination may be caused by the accumulation of holes. Moreover, under these conditions some semiconductors such as sulfides are unstable. For this reason, many studies employ additives referred to as hole scavengers to prevent these phenomena. For this purpose, researchers use reductants (organic or inorganic reducing agents as sacrificial electron donors) for donating electrons, such as triethylamine (TEA),[236] tertiary amines, especially triethanol amine (TEOA),[262] dimethylformamide and alcohols, such as methanol, ethanol, and propan-2-ol. Liu et al. carried out experiments in solvents such as water, acetonitrile, propan-2-ol, and dichloromethane with various dielectric constants. The fraction of formate increases and that of carbon monoxide decreases with the increase in the dielectric constant of the solvent. If a high dielectric-constant solvent is used, the CO_2^- anion radicals are stabilized by the solvent, resulting in weak interactions with the photocatalytic surface. Other suggested hole scavengers include ascorbic acid,[263] EDTA,[264] and Na_2SO_3.[265] Further experimental data indicate that the methanol yield increases on adding NaOH in solution.[214,215] Obviously, the caustic solution dissolves more CO_2 than pure water does and OH^- in aqueous solution serves as a strong hole scavenger. Bicarbonate and carbonate anions were also suggested to act in this role in solution.[67] The photocatalytic reaction can also be performed in gas phase using sacrificial electron donors such as H_2S,[39] H_2,[266-268] CH_3OH[234,269] and CH_4.[270,271]

7. Challenges and Prospects

The energy shortage and environmental problems are becoming the worldwide issues and the great challenges for human in the 21st century. The solar-driven transformation of CO_2 to hydrocarbons is a promising approach to regulate both the predicaments, and the potential rewards are enormous. At present, the conversion efficiency (the formation rates of products) rarely exceeds tens of $\mu mol g^{-1} h^{-1}$, indicating that the efficiency of the process is much lower than that of natural photosynthesis.

Recent developments are concentrated on new photocatalytic materials (e.g. oxynitrides of metals with d^0 electronic configurations or oxides of d^{10} metals) and on new nanoscale structures that offer a large surface area, improved charge separation, and directional electron transfers. Though great progress has been achieved toward photocatalytic conversion of CO_2 using sunlight, further effort is required for increasing sunlight-to-fuel photoconversion efficiencies.

Of particular interest are the approaches to overcome the barrier associated with the activation of a CO_2 molecule towards the first one-electron reduction, which requires a deep understanding of processes that occur on the surface of photocatalysts. Another important aspect on the fundamental understanding of the process includes charge-carrier dynamics within the semiconductors and at their interfaces. Ultrathin nanostructures were developed to facilitate the charge-carrier transport to surface reaction sites. Spatial separation of photoexcited electrons and holes can also be achieved by the design of co-catalysts, Z-figure or heterostuctures of coupling different semiconductors with properly aligned band structures. Significantly increasing the lifetime of the charge-separated state can result in a great improvement of the efficiency. Furthermore, various substrate materials have been used to provide high surface area and the catalysts' immobilization. Recently, different types of sensitization techniques, like dye sensitization, enzyme-based sensitization, QDs sensitization, and phthalocyanine sensitization, are incorporated to improve the photocatalytic performance under visible-light irradiation. The selectivity of products not only depends on the catalysts' compositions but also on the choice of reductant and the solvent used in the catalytic system. At present, many studies are focused on the half-reactions of the CO_2 reduction with the presence of sacrificial electron donors such as water, and organic reagents. However, full cycle systems, with a stronger emphasis on the oxidation half-reaction (e.g. generation of oxygen or hydrogen peroxide), will be necessary to allow for long-term studies. Other factors are definitely considerable for evaluating photocatalytic performance on CO_2 reduction such as catalysts dosage, reactants ratio, reaction temperature, time, system pressure, pH, light intensity and wavelength.

An ideal high-efficiency process is also necessary to be founded on photoactive materials made of cheap, earth abundant, non-toxic, light-stable, scalable and low cost materials. The heterogeneous photocatalytic reduction of CO_2 on semiconductors is not yet ready to be implemented in real-life solar fuel applications. However, with the current pace of developments in the field, the emerging understanding of the mechanism, as well as novel materials should bring the promise held by this process much closer to fulfillment in the near future.

References

1. S. C. Roy, O. K. Varghese, M. Paulose and C. A. Grimes. *ACS Nano* **4**, 2010, 1259–1278.
2. P. Biswas, W.-N. Wang and W.-J. An. *Frontiers of Environmental Science & Engineering in China* **5**, 201, 299–312.

3. E. J. Maginn. *The Journal of Physical Chemistry Letters* **1**, 2010, 3478–3479.
4. M. Aresta and A. Dibenedetto. *Dalton Transactions*, 2007, 2975–2992.
5. J. P. Smol, *Nature* **483**, 2012, S12–S15.
6. P. N. Pearson and M. R. Palmer. *Nature* **406**, 2000, 695–699.
7. H. Khatib. *Energy Policy* **48**, 2012,737–743.
8. I. Omae. *Coordin. Chem. Rev.* **256**, 2012, 1384–1405.
9. K. F. Li, X. Q. An, K. H. Park, M. Khraisheh and J. W. Tang. *Catal. Today* **224**, 2014, 3–12.
10. S. Dutta, A. Bhaumik and K. C. W. Wu. *Energy & Environmental Science* **7**, 2014, 3574–3592.
11. D. Y. C. Leung, G. Caramanna and M. M. Maroto-Valer. *Renew. Sust. Energ. Rev.* **39**, 2014, 426–443.
12. C. D. Windle and R. N. Perutz. *Coordin. Chem. Rev.* **256**, 2012, 2562–2570.
13. S. Das and W. Daud, *Renew. Sust. Energ. Rev.* **39**, 2014, 765–805.
14. M. Halbwachs and J.-C. Sabroux. *Science* **292**, 2001, 438.
15. S. C. Roy, O. K. Varghese, M. Paulose and C. A. Grimes. *ACS Nano* **4**, 2010, 1259–1278.
16. N. D. McDaniel and S. Bernhard. *Dalton Transactions* **39**, 2010, 10021–10030.
17. G. A. Olah, G. S. Prakash and A. Goeppert, *J. Am. Chem. Soc.* **133**, 2011, 12881–12898.
18. G. A. Olah, A. Goeppert and G. S. Prakash. *J. Org. Chem.* **74**, 2008, 487–498.
19. Y. Sharma, B. Singh and S. Upadhyay, *Fuel* **87**, 2008, 2355–2373.
20. A. Taheri Najafabadi. *Int. J. Energ. Res.* **37**, 2013, 485–499.
21. C. Costentin, M. Robert and J.-M. Savéant, *Chem. Soc. Rev.* **42**, 2013, 2423–2436.
22. Y. Chisti. *Biotechnol. Adv.* **25**, 2007, 294–306.
23. Y. Sharma, B. Singh and S. Upadhyay. *Fuel* **87**, 2008, 2355–2373.
24. N. S. Lewis. *MRS Bull.* **32**, 2007, 808–820.
25. G. Centi and S. Perathoner. *Chem. Sus. Chem.* 3, 2010, 195–208.
26. Y. Hou, B. L. Abrams, P. C. Vesborg, M. E. Björketun, K. Herbst, L. Bech, A. M. Setti, C. D. Damsgaard, T. Pedersen and O. Hansen. *Nat. Mater.* **10**, 2011, 434–438.
27. H. Zhou, X. Li, T. Fan, F. E. Osterloh, J. Ding, E. M. Sabio, D. Zhang and Q. Guo. *Adv. Mater.* **22**, 2010, 951–956.
28. M. Giampietro, S. Ulgiati and D. Pimentel. *BioScience*, 587–600, 1997.
29. L. R. Martin, *Sol. Energy* **24**, 1980, 271–277.

30. M. Galvez, P. Loutzenhiser, I. Hischier and A. Steinfeld. *Energ. Fuel* **22**, 2008, 3544–3550.
31. W. C. Chueh, C. Falter, M. Abbott, D. Scipio, P. Furler, S. M. Haile and A. Steinfeld. *Science* **330**, 2010, 1797–1801.
32. P. Furler, J. R. Scheffe and A. Steinfeld. *Energy & Environmental Science* **5**, 2012, 6098–6103.
33. H. Shibata, J. A. Moulijn and G. Mul. *Catal. Lett.* **123**, 2008, 186–192.
34. G. Centi, S. Perathoner and Z. S. Rak. *Appl. Catal. B.* **41**, 2003, 143–155.
35. C. Sanchez-Sanchez, V. Montiel, D. Tryk, A. Aldaz and A. Fujishima. *Pure Appl. Chem.* **73**, 2001, 1917–1927.
36. M. Halmann. *Nature* **275**, 1978, 115–116.
37. V. P. Indrakanti, J. D. Kubicki and H. H. Schobert, *Energy & Environmental Science* **2**, 2009, 745–758.
38. J. C. Wu. *Catal. Surv. Asia* **13**, 2009, 30–40.
39. P. Usubharatana, D. McMartin, A. Veawab and P. Tontiwachwuthikul. *Ind. Eng. Chem. Res.* **45**, 2006, 2558–2568.
40. M. Kitano, M. Matsuoka, M. Ueshima and M. Anpo. *Appl. Catal. A.* **325**, 2007, 1–14.
41. M. R. Hofmann, S. T. Martin, W. Choi and D. W. Bahnemann, *Chem. Rev.*, **95**, 69–96.
42. K. Demeestere, J. Dewulf and H. Van Langenhove. *Critical Reviews in Environmental Science and Technology* **37**, 2007, 489–538.
43. A. L. Linsebigler, G. Lu and J. T. Yates Jr. *Chem. Rev.* **95**, 1995, 735–758.
44. H. Kisch. *Angew. Chem. Int. Edit.* **52**, 2013, 812–847.
45. H. Kisch. *Angew. Chem. Int. Edit.* **125**, 2013, 842–879.
46. D. Uner, M. M. Oymak and B. Ipek. *Int. J. Global Warming* **3**, 2011, 142–162.
47. P. V. Kamat. *J. Phys. Chem. Letters* **3**, 2012, 663–672.
48. A. Fujishima, X. Zhang and D. A. Tryk. *Surface Science Reports* **63**, 2008, 515–582.
49. J.-M. Herrmann. *Appl. Catal. B.* **99**, 2010, 461–468.
50. B. Ohtani. *Chem. Lett.* **37**, 2008, 216–229.
51. A. J. Bard and M. A. Fox. *Accounts Chem. Res.* **28**, 1995, 141–145.
52. S. Styring. *Faraday Discuss.* **155**, 2012, 357–376.
53. D. G. Nocera. *Accounts Chem. Res.* **45**, 2012, 767–776.
54. K. Maeda and K. Domen. *J. Phys. Chem. Lett.* **1**, 2010, 2655–2661.
55. K. J. Young, L. A. Martini, R. L. Milot, R. C. Snoeberger, V. S. Batista, C. A. Schmuttenmaer, R. H. Crabtree and G. W. Brudvig. *Coordin. Chem. Rev.* **256**, 2012, 2503–2520.

56. T. Inoue, A. Fujishima, S. Konishi and K. Honda. *Nature* **277**, 1979, 637–638.
57. G. Dey, *J. Natural Gas Chemistry* **16**, 2007, 217–226.
58. A. Dhakshinamoorthy, S. Navalon, A. Corma and H. Garcia. *Energy Environ. Science* **5**, 2012, 9217–9233.
59. G. Liu, N. Hoivik, K. Wang and H. Jakobsen. *Sol. Energ. Mat. Sol. C.* **105**, 2012, 53–68.
60. A. Paul, D. Connolly, M. Schulz, M. T. Pryce and J. G. Vos. *Inorg. Chem.* **51**, 2012, 1977–1979.
61. K. Schouten, Y. Kwon, C. Van der Ham, Z. Qin and M. Koper. *Chemical Science* **2**, 2011, 1902–1909.
62. T. Yui, A. Kan, C. Saitoh, K. Koike, T. Ibusuki and O. Ishitani. *ACS Appl. Mater. Interfaces* **3**, 2011, 2594–2600.
63. J. Rasko and F. Solymosi. *J. Phys. Chem.* **98**, 1994, 7147–7152.
64. W. Lin and H. Frei. *J. Am. Chem. Soc.* **127**, 2005, 1610–1611.
65. C.-C. Yang, Y.-H. Yu, B. van der Linden, J. C. S. Wu and G. Mul. *J. Am. Chem. Soc.* **132**, 2010, 8398–8406.
66. M. Anpo and K. Chiba. *J. Mol. Catal.* **74**, 1992, 207–212.
67. N. M. Dimitrijevic, B. K. Vijayan, O. G. Poluektov, T. Rajh, K. A. Gray, H. He and P. Zapol. *J. Am. Chem. Soc.* **133**, 2011, 3964–3971.
68. N. M. Dimitrijevic, I. A. Shkrob, D. J. Gosztola and T. Rajh. *J. Phys. Chem. C* **116**, 2011, 878–885.
69. K. Tanaka, K. Miyahara and I. Toyoshima, *J. Phys. Chem.* **88**, 1984, 3504–3508.
70. J. Lee, D. C. Sorescu and X. Deng. *J. Am. Chem. Soc.* **133**, 2011, 10066–10069.
71. A. Kubacka, M. Fernández-García and G. Colón. *Chem. Rev.* **112**, 2011, 1555–1614.
72. H. He, P. Zapol and L. A. Curtiss. *Energy Environ. Science* **5**, 2012, 6196–6205.
73. L. Liu, W. Fan, X. Zhao, H. Sun, P. Li and L. Sun. *Langmuir* **28**, 2012, 10415–10424.
74. S. C. Roy, O. K. Varghese, M. Paulose and C. A. Grimes. *ACS Nano* **4**, 2010, 1259–1278.
75. V. P. Indrakanti, H. H. Schobert and J. D. Kubicki. *Energ. Fuel* **23**, 2009, 5247–5256.
76. S. S. Tan, L. Zou and E. Hu. *Catal. Today* **131**, 2008, 125–129.
77. A. Fujishima and K. Honda, *Nature*, 1972, 37–38.
78. A. Hagfeldt, G. Boschloo, L. Sun, L. Kloo and H. Pettersson. *Chem. Rev.* **110**, 2010, 6595–6663.

79. B. O'regan and M. Grätzel. *Nature*, **353**, 1991, 737–740.
80. H. J. Snaith and L. Schmidt-Mende. *Adv. Mater.* **19**, 2007, 3187–3200.
81. D. Cummins, G. Boschloo, M. Ryan, D. Corr, S. N. Rao and D. Fitzmaurice. *J. Phys. Chem. B* **104**, 2000, 11449–11459.
82. X. Chen and S. S. Mao. *Chem. Rev.* **107**, 2007, 2891–2959.
83. T. Tachikawa, M. Fujitsuka and T. Majima. *J. Phys. Chem. C* **111**, 2007, 5259–5275.
84. T. Ohno, K. Sarukawa, K. Tokieda and M. Matsumura. *J. Catal.* **203**, 2001, 82–86.
85. Z. Zhang, J. Lee, J. T. Yates, R. Bechstein, E. Lira, J. Ø. Hansen, S. Wendt and F. Besenbacher. *J. Phys. Chem. C* **114**, 2010, 3059–3062.
86. J. Zhang, Q. Xu, Z. Feng, M. Li and C. Li. *Angew. Chem. Int. Edit.* **47**, 2008, 1766–1769.
87. T. Kawahara, Y. Konishi, H. Tada, N. Tohge, J. Nishii and S. Ito. *Angew. Chem. Int. Edit.* **41**, 2002, 2811–2813.
88. J. Zhang, Q. Xu, Z. Feng, M. Li and C. Li. *Angew. Chemie. Int. Edit.* **120**, 2008, 1790–1793.
89. S. Kaneco, H. Kurimoto, K. Ohta, T. Mizuno and A. Saji. *J. Photoch. Photobio. A.* **109**, 1997, 59–63.
90. K. Kočí, L. Obalová, L. Matějová, D. Plachá, Z. Lacný, J. Jirkovský and O. Šolcová. *Appl. Catal. B.* **89**, 2009, 494–502.
91. P. A. Sant and P. V. Kamat. *Phys. Chem. Chem. Phys.* **4**, 2002, 198–203.
92. D. Uner, M. M. Oymak and B. Ipek. *Int. J. Global Warming* **3**, 2011, 142–162.
93. K. Kočí, L. Obalová, L. Matějová, D. Plachá, Z. Lacný, J. Jirkovský and O. Šolcová. *Appl. Catal. B.* **89**, 2009, 494–502.
94. C. B. Almquist and P. Biswas. *J. Catal.* **212**, 2002, 145–156.
95. Q.-H. Zhang, W.-D. Han, Y.-J. Hong and J.-G. Yu. *Catal. Today* **148**, 2009, 335–340.
96. J. Qu, X. Zhang, Y. Wang and C. Xie. *Electrochim. Acta* **50**, 2005, 3576–3580.
97. K. Raja, Y. Smith, N. Kondamudi, A. Manivannan, M. Misra and V. R. Subramanian. *Electrochem. Solid St.* **14**, 2011, F5–F8.
98. G. Rothenberger, D. Fitzmaurice and M. Graetzel. *J. Phys. Chem.* **96**, 1992, 5983–5986.
99. N. Serpone, D. Lawless and R. Khairutdinov. *J. Phys. Chem.* **99**, 1995, 16646–16654.
100. D. C. Hurum, A. G. Agrios, K. A. Gray, T. Rajh and M. C. Thurnauer. *J. Physical Chemistry B*, **107**, 2003, 4545–4549.

101. G. Li and K. A. Gray. *Chemical physics*, **339**, 2007, 173–187.
102. H. He, C. Liu, K. D. Dubois, T. Jin, M. E. Louis and G. Li. *Industrial & Engineering Chemistry Research*, **51**, 2012, 11841–11849.
103. T. Ohno, K. Sarukawa, K. Tokieda and M. Matsumura. *J. Catal.* **203**, 2001, 82–86.
104. T. Ohno, K. Tokieda, S. Higashida and M. Matsumura. *Appl. Catal. A* **244**, 2003, 383–391.
105. J. Zhang, Q. Xu, Z. Feng, M. Li and C. Li. *Angew. Chem. Int. Edit.* **47**, 2008, 1766–1769.
106. R. Su, R. Bechstein, L. Sø, R. T. Vang, M. Sillassen, B. R. Esbjörnsson, A. Palmqvist and F. Besenbacher. *J. Phys. Chem. C* **115**, 2011, 24287–24292.
107. R. I. Bickley, T. Gonzalez-Carreno, J. S. Lees, L. Palmisano and R. J. Tilley. *J. Solid State Chem.* **92**, 1991, 178–190.
108. D. C. Hurum, A. G. Agrios, K. A. Gray, T. Rajh and M. C. Thurnauer. *J. Phys. Chem. B* **107**, 2003, 4545–4549.
109. R. G. Nair, S. Paul and S. Samdarshi. *Sol. Energ. Mat. Sol. C.* **95**, 2011, 1901–1907.
110. M. Anpo, H. Yamashita, Y. Ichihashi and S. Ehara. *J. Electroanal. Chem.* **396**, 1995, 21–26.
111. M. Nowotny, L. Sheppard, T. Bak and J. Nowotny. *J. Phys. Chem. C*, **112**, 2008, 5275–5300.
112. X. Chen, L. Liu, Y. Y. Peter and S. S. Mao. *Science* **331**, 2011, 746–750.
113. S. Hoang, S. P. Berglund, N. T. Hahn, A. J. Bard and C. B. Mullins. *J. Am. Chem. Soc.* **134**, 2012, 3659–3662.
114. M. Xing, W. Fang, M. Nasir, Y. Ma, J. Zhang and M. Anpo. *J. Catal.* **297**, 2013, 236–243.
115. P. A. DeSario, L. Chen, M. E. Graham and K. A. Gray. *J. Vac. Sci. Tech. A* **29**, 2011, 031508.
116. J. Nowotny, T. Bak, M. Nowotny and L. Sheppard. *J. Phys. Chem. B* **110**, 2006, 18492–18495.
117. D. Acharya, N. Camillone III and P. Sutter. *J. Phys. Chem. C* **115**, 2011, 12095–12105.
118. Z. Lin, A. Orlov, R. M. Lambert and M. C. Payne. *J. Phys. Chem. B* **109**, 2005, 20948–20952.
119. M. Xing, W. Fang, M. Nasir, Y. Ma, J. Zhang and M. Anpo. *J. Catal.* **297**, 2013, 236–243.
120. V. Jeyalakshmi, K. Rajalakshmi, R. Mahalakshmy, K. Krishnamurthy and B. Viswanathan. *Res. Chem. Intermediat.* **39**, 2013, 2565–2602.

121. R. K. de Richter, T. Z. Ming and S. Caillol. *Renew. Sust. Energ. Rev.* **19**, 2013, 82–106.
122. P. D. Tran, L. H. Wong, J. Barber and J. S. C. Loo. *Energy & Environmental Science* **5**, 2012, 5902–5918.
123. Z. Wang, J. Lu, A. Peer and M. Buss, in *Haptics: Generating and perceiving tangible sensations*, 2010, pp. 172–177, Springer.
124. M. H. Huang, S. Mao, H. Feick, H. Yan, Y. Wu, H. Kind, E. Weber, R. Russo and P. Yang. *Science* **292**, 2001, 1897–1899.
125. Y. Liu, B. Huang, Y. Dai, X. Zhang, X. Qin, M. Jiang and M.-H. Whangbo. *Catal. Commun.* **11**, 2009, 210–213.
126. J. Yu and X. Yu. *Environ. Sci. Technol.* **42**, 2008, 4902–4907.
127. K. Teramura, H. Tsuneoka, T. Shishido and T. Tanaka. *Chem. Phys. Lett.* **467**, 2008, 191–194.
128. Y. Inoue. *Energy & Environmental Science* **2**, 2009, 364–386.
129. X. Chen, S. Shen, L. Guo and S. S. Mao, *Chem. Rev.* **110**, 2010, 6503–6570.
130. K. Maeda, K. Teramura, N. Saito, Y. Inoue, H. Kobayashi and K. Domen. *Pure Appl. Chem.* **78**, 2006, 2267–2276.
131. H.-a. Park, J. H. Choi, K. M. Choi, D. K. Lee and J. K. Kang. *J. Mater. Chem.* **22**, 2012, 5304–5307.
132. M. D. Doherty, D. C. Grills, J. T. Muckerman, D. E. Polyansky and E. Fujita. *Coordin. Chem. Rev.* **254**, 2010, 2472–2482.
133. H. Tsuneoka, K. Teramura, T. Shishido and T. Tanaka. *J. Phys. Chem. C*, **114**, 2010, 8892–8898.
134. Y.-X. Pan, C.-J. Liu, D. Mei and Q. Ge. *Langmuir* **26**, 2010, 5551–5558.
135. M. Calatayud, S. E. Collins, M. A. Baltanas and A. L. Bonivardi. *Phys. Chem. Chem. Phys.* **11**, 2009, 1397–1405.
136. S. E. Collins, M. A. Baltanás and A. L. Bonivardi. *J. Phys. Chem. B* **110**, 2006, 5498–5507.
137. P.-W. Pan and Y.-W. Chen. *Catal. Commun.* **8**, 2007, 1546–1549.
138. P.-Y. Liou, S.-C. Chen, J. C. S. Wu, D. Liu, S. Mackintosh, M. Maroto-Valer and R. Linforth. *Energy & Environmental Science*, **4**, 2011, 1487–1494.
139. D.-S. Lee, H.-J. Chen and Y.-W. Chen. *J. Phys. Chem. Solids* **73**, 2012, 661–669.
140. X. Li, H. Pan, W. Li and Z. Zhuang. *Appl. Catal. A* **413**, 2012, 103–108.
141. Y. Zhou, Z. Tian, Z. Zhao, Q. Liu, J. Kou, X. Chen, J. Gao, S. Yan and Z. Zou. *ACS Appl. Mater. Interfaces* **3**, 2011, 3594–3601.

142. P. Li, S. Ouyang, G. Xi, T. Kako and J. Ye. *J. Phys. Chem. C* **116**, 2012, 7621–7628.
143. K. Iizuka, T. Wato, Y. Miseki, K. Saito and A. Kudo. *J. Am. Chem. Soc.* **133**, 2011, 20863–20868.
144. T. Omata, N. Ueda, K. Ueda and H. Kawazoe. *Appl. Phys. Lett.* **64**, 1994, 1077–1078.
145. X. Chen, H. Xue, Z. Li, L. Wu, X. Wang and X. Fu. *J. Phys. Chem. C* **112**, 2008, 20393–20397.
146. S. E. Collins, M. A. Baltanás and A. L. Bonivardi. *J. Phys. Chem. B* **110**, 2006, 5498–5507.
147. S. C. Yan, S. X. Ouyang, J. Gao, M. Yang, J. Y. Feng, X. X. Fan, L. J. Wan, Z. S. Li, J. H. Ye, Y. Zhou and Z. G. Zou. *Angew. Chem. Int. Edit.* **122**, 2010, 6544–6548.
148. S. C. Yan, S. X. Ouyang, J. Gao, M. Yang, J. Y. Feng, X. X. Fan, L. J. Wan, Z. S. Li, J. H. Ye, Y. Zhou and Z. G. Zou. *Angew. Chem. Int. Edit.* **49**, 2010, 6400–6404.
149. J. Yu and A. Kudo. *Adv. Func. Mater.* **16**, 2006, 2163–2169.
150. X. Li, J. Chen, H. Li, J. Li, Y. Xu, Y. Liu and J. Zhou. *Journal of Natural Gas Chemistry*, **20**, 2011, 413–417.
151. J. Mao, T. Peng, X. Zhang, K. Li, L. Ye and L. Zan. *Catal. Sci. Technol.* **3**, 2013, 1253–1260.
152. K. Zhang and L. Guo. *Catal. Sci. Technol.* **3**, 2013, 1672–1690.
153. F. R. F. Fan, P. Leempoel and A. J. Bard, *J. Electrochem. Soc.* **130**, 1983, 1866–1875.
154. H. Fujiwara, H. Hosokawa, K. Murakoshi, Y. Wada, S. Yanagida, T. Okada and H. Kobayashi. *J. Phys. Chem. B* **101**, 1997, 8270–8278.
155. A. Henglein, M. Gutiérrez and C. H. Fischer. *Berichte der Bunsengesellschaft für physikalische Chemie*, **88**, 1984, 170–175.
156. M. Kanemoto, T. Shiragami, C. Pac and S. Yanagida. *J. Phys. Chem.* **96**, 1992, 3521–3526.
157. H. Inoue, H. Moriwaki, K. Maeda and H. Yoneyama. *J. Photochem. Photobio. A* **86**, 1995, 191–196.
158. H. Yoneyama. *Catal. Today* **39**, 1997, 169–175.
159. R. Vogel, P. Hoyer and H. Weller. *J. Phys. Chem.* **98**, 1994, 3183–3188.
160. D. Meissner, R. Memming and B. Kastening. *J. Phys. Chem.* **92**, 1988, 3476–3483.
161. B.-J. Liu, T. Torimoto and H. Yoneyama. *J. Photoch. Photobio. A* **113**, 1998, 93–97.

162. X. V. Zhang, S. T. Martin, C. M. Friend, M. A. A. Schoonen and H. D. Holland. *J. Am. Chem. Soc.* **126**, 2004, 11247–11253.
163. S. Kaneco, H. Katsumata, T. Suzuki and K. Ohta. *Chem. Eng. J.* **116**, 2006, 227–231.
164. E. E. Barton, D. M. Rampulla and A. B. Bocarsly. *J. Am. Chem. Soc.* **130**, 2008, 6342–6344.
165. S. N. Habisreutinger, L. Schmidt-Mende and J. K. Stolarczyk. *Angew. Chem. Int. Edit.* **52**, 2013, 7372–7408.
166. B. A. Parkinson and P. F. Weaver. *Nature* **309**, 1984, 148–149.
167. T. Arai, S. Sato, K. Uemura, T. Morikawa, T. Kajino and T. Motohiro, *Chem. Commun.* **46**, 2010, 6944–6946.
168. S. Sato, T. Arai, T. Morikawa, K. Uemura, T. M. Suzuki, H. Tanaka and T. Kajino. *J. Am. Chem. Soc.* **133**, 2011, 15240–15243.
169. S. Yan, H. Yu, N. Wang, Z. Li and Z. Zou. *Chem. Commun.* **48**, 2012, 1048–1050.
170. Q. Liu, Y. Zhou, Z. Tian, X. Chen, J. Gao and Z. Zou. *J. Mater. Chem.* **22**, 2012, 2033–2038.
171. N. Zhang, S. Ouyang, T. Kako and J. Ye. *Chem. Commun.* **48**, 2012, 1269–1271.
172. S. Yan, J. Wang, H. Gao, N. Wang, H. Yu, Z. Li, Y. Zhou and Z. Zou. *Advanced Functional Materials*, **23**, 2013, 1839–1845.
173. I. Tsuji, H. Kato and A. Kudo. *Chem. Mater.* **18**, 2006, 1969–1975.
174. J.-Y. Liu, B. Garg and Y.-C. Ling. *Green Chem.* **13**, 2011, 2029–2031.
175. C. Wang, Z. Xie, K. E. deKrafft and W. Lin. *J. Am. Chem. Soc.* **133**, 2011, 13445–13454.
176. J.-L. Wang, C. Wang and W. Lin. *ACS Catalysis*, **2**, 2012, 2630–2640.
177. Z.-D. Meng, L. Zhu, J.-G. Choi, M.-L. Chen and W.-C. Oh. *J. Mater. Chem.* **21**, 2011, 7596–7603.
178. K.-D. Kim, N. K. Dey, H. O. Seo, Y. D. Kim, D. C. Lim and M. Lee. *Appl. Catal. A* **408**, 2011, 148–155.
179. K. Woan, G. Pyrgiotakis and W. Sigmund. *Adv. Mater.* **21**, 2009, 2233–2239.
180. H. Zhang, X. Lv, Y. Li, Y. Wang and J. Li. *ACS Nano*, **4**, 2009, 380–386.
181. S. Liu, H. Sun, S. Liu and S. Wang. *Chem. Eng. J.* **214**, 2013, 298–303.
182. J. Yu, B. Yang and B. Cheng. *Nanoscale* **4**, 2012, 2670–2677.
183. C. Martínez, M. Canle, L. M. I. Fernández, J. A. Santaballa and J. Faria. *Appl. Catal.* **102**, 2011, 563–571.
184. X.-H. Xia, Z.-J. Jia, Y. Yu, Y. Liang, Z. Wang and L.-L. Ma. *Carbon* **45**, 2007, 717–721.

185. W.-J. Ong, M. M. Gui, S.-P. Chai and A. R. Mohamed. *RSC Advances* **3**, 2013, 4505–4509.
186. Q.-P. Luo, X.-Y. Yu, B.-X. Lei, H.-Y. Chen, D.-B. Kuang and C.-Y. Su. *J. Phys. Chem. C* **116**, 2012, 8111–8117.
187. Z. Xiong, L. L. Zhang, J. Ma and X. S. Zhao. *Chem. Commun.* **46**, 2010, 6099–6101.
188. W. Tu, Y. Zhou, Q. Liu, S. Yan, S. Bao, X. Wang, M. Xiao and Z. Zou. *Advanced Functional Materials*, **23**, 2013, 1743–1749.
189. L. L. Zhang, Z. Xiong and X. S. Zhao. *ACS Nano*, **4**, 2010, 7030–7036.
190. G. Liao, S. Chen, X. Quan, H. Yu and H. Zhao. *J. Mater. Chem.* **22**, 2012, 2721–2726.
191. Y. Xu, H. Xu, L. Wang, J. Yan, H. Li, Y. Song, L. Huang and G. Cai. *Dalton Transactions* **42**, 2013, 7604–7613.
192. H. Tong, S. Ouyang, Y. Bi, N. Umezawa, M. Oshikiri and J. Ye. *Adv. Mater.* **24**, 2012, 229–251.
193. A. Kudo and Y. Miseki. *Chem. Soc. Rev.* **38**, 2009, 253–278.
194. I. Nakamura, N. Negishi, S. Kutsuna, T. Ihara, S. Sugihara and K. Takeuchi. *J. of Mol. Catal. A* **161**, 2000, 205–212.
195. H. Irie, Y. Watanabe and K. Hashimoto. *Chem. Lett.* **32**, 2003, 772–773.
196. Y. Gai, J. Li, S.-S. Li, J.-B. Xia and S.-H. Wei. *Phys. Rev. Lett.* **102**, 2009.
197. R. Long and N. J. English. *Appl. Phys. Lett.* **94**, 2009, 132102.
198. K. Sayama and H. Arakawa. *J. Phys. Chem.* **97**, 1993, 531–533.
199. X. Li, Z. Zhuang, W. Li and H. Pan. *Applied Catalysis A* **429–430**, 2012, 31–38.
200. S. Sato, T. Morikawa, S. Saeki, T. Kajino and T. Motohiro. *Angew. Chem. Int. Edit.* **49**, 2010, 5101–5105.
201. T. M. Suzuki, H. Tanaka, T. Morikawa, M. Iwaki, S. Sato, S. Saeki, M. Inoue, T. Kajino and T. Motohiro. *Chem. Commun.* **47**, 2011, 8673–8675.
202. C.-W. Tsai, H. M. Chen, R.-S. Liu, K. Asakura and T.-S. Chan. *J. Phys. Chem. C* **115**, 2011, 10180–10186.
203. H. Irie, S. Washizuka and K. Hashimoto, *Thin Solid Films.* **510**, 2006, 21–25.
204. T. Ohno, M. Akiyoshi, T. Umebayashi, K. Asai, T. Mitsui and M. Matsumura. *Appl. Catal. A* **265**, 2004, 115–121.
205. D. Li, H. Haneda, S. Hishita, N. Ohashi and N. K. Labhsetwar. *J. Fluorine Chem.* **126**, 2005, 69–77.
206. Z. He, L. Xie, S. Song, C. Wang, J. Tu, F. Hong, Q. Liu, J. Chen and X. Xu. *J. of Mol. Catal. A* **319**, 2010, 78–84.

207. Q. Zhang, Y. Li, E. A. Ackerman, M. Gajdardziska-Josifovska and H. Li. *Appl. Catal. A* **400**, 2011, 195–202.
208. S.-I. In, D. D. Vaughn and R. E. Schaak. *Angew. Chem. Int. Edit.* **51**, 2012, 3915–3918.
209. M. Ni, M. K. H. Leung, D. Y. C. Leung and K. Sumathy. *Renew. Sust. Energ. Rev.* **11**, 2007, 401–425.
210. N. Semagina and L. Kiwi-Minsker. *Catal. Lett.* **127**, 2009, 334–338.
211. V. M. Daskalaki and D. I. Kondarides. *Catal. Today* **144**, 2009, 75–80.
212. O. Ishitani. *J. Photochem. Photobio. A* **72**, 1993, 269–271.
213. F. Solymosi, I. Tombácz and J. Koszta. *J. Catal.* **95**, 1985, 578–586.
214. I. H. Tseng, J. C. S. Wu and H.-Y. Chou. *J. Catal.* **221**, 2004, 432–440.
215. I. H. Tseng, W.-C. Chang and J. C. S. Wu. *Appl. Catal. B* **37**, 2002, 37–48.
216. K. Adachi, K. Ohta and T. Mizuno. *Sol. Energy* **53**, 1994, 187–190.
217. K. Kočí, K. Matějů, L. Obalová, S. Krejčíková, Z. Lacný, D. Plachá, L. Čapek, A. Hospodková and O. Šolcová. *Appl. Catal. B* **96**, 2010, 239–244.
218. S. Krejčíková, L. Matějová, K. Kočí, L. Obalová, Z. Matěj, L. Čapek and O. Šolcová. *Appl. Catal. B* **111–112**, 2012, 119–125.
219. P. Ji, M. Takeuchi, T.-M. Cuong, J. Zhang, M. Matsuoka and M. Anpo. *Res. Chem. Intermediat.* **36**, 2010, 327–347.
220. M. Alvaro, C. Aprile, B. Ferrer, F. Sastre and H. Garcia. *Dalton Transactions*, 2009, 7437–7444.
221. C. Aprile, M. A. Herranz, E. Carbonell, H. Garcia and N. Martin. *Dalton Transactions*, 2009, 134–139.
222. G. Guan, T. Kida and A. Yoshida. *Appl. Catal. B* **41**, 2003, 387–396.
223. A. Nishimura, G. Mitsui, M. Hirota and E. Hu. *Int. J. Chem. Eng.* 2010.
224. R. Dholam, N. Patel, M. Adami and A. Miotello. *Int. J. Hydrogen Energ.* **34**, 2009, 5337–5346.
225. L. Pan, J.-J. Zou, X. Zhang and L. Wang. *Ind. Eng. Chem. Res.* **49**, 2010, 309103, 8526–8531.
226. T.-V. Nguyen and J. C. S. Wu. *Appl. Catal. A* **335**, 2008, 112–120.
227. T.-V. Nguyen and J. C. S. Wu. *Sol. Energ. Mat. Sol. Cells* **92**, 2008, 864–872.
228. Q. Zhai, S. Xie, W. Fan, Q. Zhang, Y. Wang, W. Deng and Y. Wang. *Angew. Chem. Int. Edit.* **125**, 2013, 5888–5891.
229. O. K. Varghese, M. Paulose, T. J. LaTempa and C. A. Grimes. *Nano Lett.* **9**, 2009, 731–737.
230. B. D. Mankidy, B. Joseph and V. K. Gupta. *Nanotechnol.* **24**, 2013.
231. M. QAMAR. *Int. J. Nanoscience* **09**, 2010, 579–583.

232. H. Zhu, N. Song, W. Rodríguez-Córdoba and T. Lian. *J. Am. Chem. Soc.* **134**, 2012, 4250–4257.
233. G. K. Mor, O. K. Varghese, R. H. T. Wilke, S. Sharma, K. Shankar, T. J. Latempa, K.-S. Choi and C. A. Grimes. *Nano Lett.* **8**, 2008, 1906–1911.
234. S. Qin, F. Xin, Y. Liu, X. Yin and W. Ma. *J. Colloid Interf. Sci.* **356**, 2011, 257–261.
235. Q. D. Truong, J.-Y. Liu, C.-C. Chung and Y.-C. Ling. *Cataly. Commun.* **19**, 2012, 85–89.
236. Slamet, H. W. Nasution, E. Purnama, S. Kosela and J. Gunlazuardi. *Catal. Commun.* **6**, 2005, 313–319.
237. H. Jüntgen. *Fuel*, **65**, 1986, 1436–1446.
238. M. Kruk, M. Jaroniec, C. H. Ko and R. Ryoo. *Chem. Mater.* **12**, 2000, 1961–1968.
239. K. Ikeue, S. Nozaki, M. Ogawa and M. Anpo. *Catal. Today* **74**, 2002, 241–248.
240. W. Lin, H. Han and H. Frei. *J. Phys. Chem. B* **108**, 2004, 18269–18273.
241. Y. Shioya, K. Ikeue, M. Ogawa and M. Anpo. *Appl. Catal. A* **254**, 2003, 251–259.
242. N. Sasirekha, S. J. S. Basha and K. Shanthi. *Appl. Catal. B* **62**, 2006, 169–180.
243. M. Anpo, H. Yamashita, Y. Ichihashi, Y. Fujii and M. Honda. *J. Physical Chemistry B* **101**, 1997, 2632–2636.
244. M. Tahir and N. S. Amin. *Appl. Catal. Enviro* **142–143**, 2013, 512–522.
245. K. Kočí, V. Matějka, P. Kovář, Z. Lacný and L. Obalová. *Catal. Today* **161**, 2011, 105–109.
246. J. Jensen, M. Mikkelsen and F. C. Krebs. *Sol. Energ. Mater. Sol. Cells* **95**, 2011, 2949–2958.
247. P. Gijsman, G. Meijers and G. Vitarelli. *Polym. Degrad. Stabil.* **65**, 1999, 433–441.
248. C. Jin, P. A. Christensen, T. A. Egerton, E. J. Lawson and J. R. White. *Polym. Degrad. Stabil.* **91**, 2006, 1086–1096.
249. X. Li, H. Liu, D. Luo, J. Li, Y. Huang, H. Li, Y. Fang, Y. Xu and L. Zhu. *Chem. Eng. J.* **180**, 2012, 151–158.
250. C. Wang, R. L. Thompson, J. Baltrus and C. Matranga. *J. Phys. Chem. Lett.* **1**, 2009, 48–53.
251. C. Wang, R. L. Thompson, P. Ohodnicki, J. Baltrus and C. Matranga. *J. Mater. Chem.* **21**, 2011, 13452–13457.
252. M. Abou Asi, C. He, M. Su, D. Xia, L. Lin, H. Deng, Y. Xiong, R. Qiu and X.-Z. Li. *Catal. Today* **175**, 2011, 256–263.

253. M. Tahir and N. S. Amin. *Energ. Convers. Manage.* **76**, 2013, 194–214.
254. W. A. Tisdale, K. J. Williams, B. A. Timp, D. J. Norris, E. S. Aydil and X.-Y. Zhu. *Science* **328**, 2010, 1543–1547.
255. T.-V. Nguyen, J. C. S. Wu and C.-H. Chiou. *Catal. Commun.* **9**, 2008, 2073–2076.
256. O. Ozcan, F. Yukruk, E. U. Akkaya and D. Uner. *Top Catal.* **44**, 2007, 523–528.
257. Y. Yang, H. Zhong and C. Tian. *Res. Chem. Inter.* **37**, 2011, 91–102.
258. Z. Zhao, J. Fan, M. Xie and Z. Wang. *J. Cleaner Production* **17**, 2009, 1025–1029.
259. Z. Zhao, J. Fan, S. Liu and Z. Wang. *Chem. Eng. J.* **151**, 2009, 134–140.
260. Z.-H. Zhao, J.-M. Fan and Z.-Z. Wang. *J. Cleaner Production* **15**, 2007, 1894–1897.
261. Q. Wang, W. Wu, J. Chen, G. Chu, K. Ma and H. Zou. *Colloid. Surface. A* **409**, 2012, 118–125.
262. I. H. Tseng and J. C. S. Wu. *Catal. Today* **97**, 2004, 113–119.
263. C. A. Craig, L. O. Spreer, J. W. Otvos and M. Calvin. *J. Phys. Chem.* **94**, 1990, 7957–7960.
264. G. Kim and W. Choi. *Appl. Catal. B* **100**, 2010, 77–83.
265. S. Liu, Z. Zhao and Z. Wang. *Photochem. Photobio. Sci.* **6**, 2007, 695–700.
266. J.-L. Li, N. Fu and G.-X. Lue. *Chin. J. Inorg. Chem.* **26**, 2010, 2175–2181.
267. Y. Kohno, H. Hayashi, S. Takenaka, T. Tanaka*, T. Funabiki and S. Yoshida. *J. Photochem. Photobio. A* **126**, 1999, 117–123.
268. S. S. Tan, L. Zou and E. Hu. *Sci. Technol. Adv. Mat.* **8**, 2007, 89–92.
269. N. Ulagappan and H. Frei. *J. Phys. Chem. A* **104**, 2000, 7834–7839.
270. D. Shi, Y. Feng and S. Zhong. *Catal. Today* **98**, 2004, 505–509.
271. T. W. Woolerton, S. Sheard, E. Reisner, E. Pierce, S. W. Ragsdale and F. A. Armstrong. *J. Am. Chem. Soc.* **132**, 2010, 2132–2133.

Index

Ag nanoparticles, 120
artificial photosynthesis, 197
Au nanoparticles, 129

black brookite TiO_2 nanoparticles, 18
black rutile TiO_2 nanoparticles, 18
black TiO_2 nanoparticles, 2
black titanium dioxide, 1

carrier injection, 120, 128
CO_2 reduction, 125
conduction band minimum, 19
core–shell structure, 5
crystalline core/disordered shell, 5
crystalline/disordered core/shell structures, 14

density functional theory, 18
doping, 2, 37
doping external elements, 266

electroless chemical etching, 60
environmental pollution removal, 1

field enhancement, 120, 124
First-principles band structure theory calculation, 19

graphene, 82, 214

hematite, 28
heteronanostructures, 36
heterostructures, 269
high-pressure hydrogenation, 3
hot carriers, 122
hot charge carriers, 160
hot electron, 129
hydrocarbons, 238
hydrogenated TiO_2 nanoparticles, 4
hydrogen evolution, 197

Landau damping, 122
lithium-ion microbatteries, 21
local electric field enhancement, 161
localized surface plasmon, 157
localized surface plasmon resonance, 119

MOFs, 261
molecular hydrogen (H_2), 3
multi-electron transfer processes, 241

nanocarbons, 263

organic synthesis, 83
organic transformations, 167
oxide, 253
oxygen evolution reaction, 28
oxygen vacancy, 2

P25 TiO_2 powders, 5
phase transformation, 5
phosphide, 253
photocatalysis, 57, 81, 117, 156
photocatalytic conversion of CO_2, 234
photocatalytic water splitting, 1, 125
photoelectrochemical, 27
photoelectrochemical water-splitting, 20
photogenerated electrons and holes, 236
photosensitization, 272
photothermal conversion, 162
plasmon-enhanced photocatalysis, 118
plasmonic catalysis, 167
plasmonic heating, 120, 137

plasmonic nanoparticles, 118
porous SiNWs, 62
proton pump, 198

rational design, 245
reaction mechanisms, 119
retinal, 199
rutile TiO_2 nanowire arrays, 6

selective, 82
self-assembly, 215
silicon nanowires, 58
solar, 118
solid solutions, 260
state-selective reactions, 144
sulfide, 253
sun light, 120, 143, 144
supercapacitors, 21
supported materials, 270

TiO_2, 203, 243
TiO_2 nanotube, 7
titanium dioxide (TiO_2), 2

valence band maximum, 19
visible-light driven catalysis, 119
visible-light-sensitive photocatalyst, 239

water splitting, 28, 131

Printed in the United States
By Bookmasters